AF893832

Renewable Energy at Home

A Hands-on Guide to Crafting Your Own Power Plant

Alex Pozhitkov and boB Gudgel

elektor
design > share > earn

This is an Elektor Publication. Elektor is the media brand of
Elektor International Media B.V.
PO Box 11, NL-6114-ZG Susteren, The Netherlands
Phone: +31 46 4389444

Declaration

British Library Cataloguing in Publication Data
A catalogue record for this book is available from the British Library

ISBN 978-3-89576-590-2 Print
ISBN 978-3-89576-591-9 eBook

Editor: Jan Buiting, MA
Prepress Production: D-Vision, Julian van den Berg
Print: Ipskamp Printing, Enschede (NL)

Contents

For Whom is This Written? **7**

Tools, Skills, and Supplies **8**

Learn CAD 8

Data Logger 8

Tools 8

Suppliers and Supplies. 8

References 8

Book Organization. 9

Abbreviations **10**

Chapter 1 • Helpful DIY Instruments. **11**

Pi-logger 11

Pyranometer. 14

Chapter 2 • Power from the Sky **25**

Theoretical Foundation. 25

Garage Powerplant 31

Mounting Solar Panels 32

On the Roof 34

From Photons to Electrons 37

Chapter 3 • Wind Energy **46**

Power in the Air 46

Turbine Setup. 49

Chapter 4 • Curious Power Sources **55**

Chapter 5 • Conduits and Batteries **60**

Conduits 60

Lead-acid Batteries 61

Literature **65**

Appendix. **67**

Program: Voltread. 67

Program: Voltlog. 70

Hybrid inverter 74

DS18B20 85
SUVT Plexiglas 86
LTC3108. 87
Index 109

For Whom is This Written?

Many books have already been written about various aspects of renewable energy, *e.g.*, economic, political, etc. In addition, there are also books for the DIYers claiming to guide creating an off-grid solar energy powered house. The book you are reading now was written while we created a garage and a deck "power plants" from scratch. We present actual steps and considerations relevant to establishing your own energy independence together with some revealing experiments along the way.

We speak to the experimenters and geeks who would like to play with their renewable energy setup, learn fundamentals experimentally and ultimately build their own unique solution. The readers are encouraged to follow the projects described here and produce their own variations specific to their locations / circumstances. Some of the experiments and projects may sound like reinvention of a wheel, however we firmly believe that it is okay to re-invent and make a couple of wheels in someone's lifetime if it provides the experience and in-depth understanding of the subject.

Finally, our intention was not to simply rephrase information from various sources to fill up the pages, but rather create a new hands-on experience that will grow on its own. Where appropriate, we provide references to published literature and stable (hopefully) web resources for additional information for those who are interested.

Tools, Skills, and Supplies

Learn CAD

Creating your own renewable energy power plant requires serious attitude and attention to details. Even if you are not an engineer, you will have to become one, at least partially. Hence, planning your design will require some CAD work. We are not talking about thousands of dollars CAD software, rather there are wonderful CAD products like DesignSpark Mechanical (DSM). There are multiple video tutorials bringing you up to speed with CAD. For any DIY engineer, CAD is simply indispensable. All designs described in this book are accompanied by DSM files.

Data Logger

We will be evaluating the performance of our solar and wind setups in terms of voltage and current output. Also, we will monitor the amount of solar light as a function of time in our specific location. Moderately sophisticated charge controllers will provide data logging, some of which will even allow online monitoring. Nevertheless, monitoring the wind turbine performance or measuring the amount of solar energy requires a separate data logger. There are professional expensive data loggers providing several channels for voltage and current logging. Here we suggest a cheaper and more creative solution based on Raspberry Pi single board computer (https://www.raspberrypi.com/). Specifically, we used Raspberry Pi Zero 2W, which is a very small single board computer with Wi-Fi. Further in the book we provided details about setting up and deploying the data logger. Before that, learn some basics of Linux!

Tools

One wise man's words: "with appropriate tools one could achieve virtually anything". We highly recommend investing in a set of tools. There is nothing more frustrating than working with inadequate equipment. Please, do not attempt assembling a project holding parts in the air or with your feet! Get yourself a drill press, hand drill with a hammer mode, impact driver, a set of "number" drill bits, tap, die, jigsaw, angle grinder, belt grinder and a circular saw. All of these can be purchased over time from a home improvement stores relatively inexpensively.

Suppliers and Supplies

Home improvement stores like Home Depot, Lowe's (USA) or Bauhaus (Germany) are filled with great items like pipes, fasteners, metal angles, sheets and rods, conduits, wires as well as tools and consumables. Besides the usual online shopping, *e.g.*, Amazon, also consider industrial suppliers such as McMaster-Carr, Grainger, Online Metals (ThyssenKrupp AG). These suppliers have an incredible assortment of fasteners, wires, raw materials, chemicals, and tools. For the electronic components, consider Digi-Key, Mouser and Conrad Elektronik.

References

There is a very useful "Pocket Reference" book (1) that provides invaluable information on fasteners and corresponding hole sizes, ampacity of wires, material properties, etc. It is best to avoid the choice of materials and fasteners based on feelings, rather than on well-established standards. These standards came about through trial and error, and in this case the re-invention of the wheel is not helpful. After all, inadequate fasteners may corrode or be not strong enough resulting in a collapse of your solar or wind energy setup.

Book Organization

The fist chapter of the book proposes two DIY instruments, which we will use while building our renewable energy system. These instruments perhaps may be replaced with the off-the shelf analogs, however building, and using our own may enrich the whole experience. The following chapter is devoted to photovoltaics, from theory to the actual garage "power plant". The fourth chapter deals with wind energy and a small wind turbine, which can be placed on the roof or deck. Another chapter describes curious power sources and suggests building a microbial cell battery based on rabbit manure. The manure battery is quite weak, however playing around with such power sources may be inspirational for curious researchers. The last chapter discusses technical aspects of wiring and battery recovery.

Abbreviations

ABS – acrylonitrile butadiene styrene
Ah – ampère-hour (capacity)
ADC – analog to digital converter
AM – air mass coefficient
FET – field effect transistor
JAE – Japan Aviation Electronics
I – current
MPP – maximum power point
PV – photovoltaic
RPM – revolutions per minute
SMPSU – switch mode power supply unit
V – voltage

Chapter 1 • Helpful DIY Instruments

Pi-logger

A data logger can be very helpful to investigate the performance of our power plant, whether it is a microbial power source, solar panels, or a wind turbine. Specifically, the data logger is used for measuring the amount of solar radiation (*i.e.*, insolation) by a pyranometer, as discussed in the photovoltaics chapter. Also, the wind turbine presented in the following chapter can be optimized based on the measurements obtained by the logger. There are many professional data loggers available on the market, which are capable of measuring voltage and current across several channels. These solutions are quite expensive and do not seem to offer a "1-wire" interface, which is used for temperature measurements among other things. Here we propose a data logger, the "Pi-logger" based on Raspberry Pi Zero 2W single board computer, featuring Wi-Fi as well as several analog and digital inputs. The schematic of the Pi-logger along with actual implementation is shown in Figure 1.

The Pi-logger runs Linux operating system (OS). Raspberry Pi web pages provide step by step instructions about the installation and configuration of the OS. Briefly, a microSD card must be loaded with a bootable Linux ISO image using the Rufus software. There is also a Raspberry Pi imager software that serves the same purpose. The Linux image may be obtained from the Ubuntu or Raspberry Pi web sites. Choose the Linux installation without the desktop because the Pi-logger works with text-based interface, while graphics pose unnecessary complications. During the installation, it is important to have the keyboard and monitor connected to see the progress, to create an admin account and provide Wi-Fi settings. After the installation, the Raspberry Pi will be used "headless", *i.e., only using the terminal software over the network. To make the system up to date, run*

```
sudo apt update
sudo full-upgrade
```

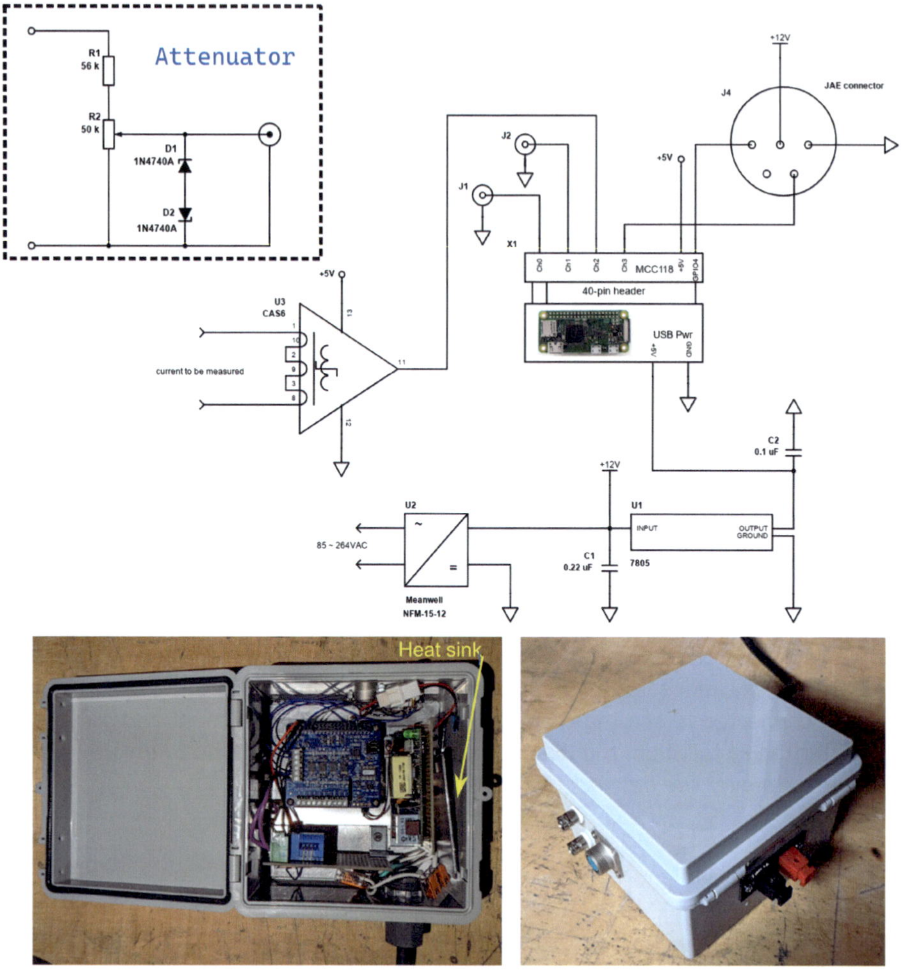

Figure 1: Pi-logger schematic (top) and actual implementation in a water-resistant enclosure. The ADC MCC 118 hat (blue PCB) is placed on top of the Raspberry Pi Zero 2W. Power is provided by a Meanwell NFM-15-12 SMPSU. A linear power regulator 7805 is outfitted with a 1.9x2.9x0.06" aluminum sheet as a heat sink. Red and black spring-loaded connectors are the input for measuring current. Inset: a schematic diagram of attenuator to expand the measurable voltage range.

The Raspberry Pi Zero 2W is sandwiched with another board, MCC 118 ADC ("ADC hat", by Measurement Computing Corporation). Mains AC voltage is conditioned using a single-board SMPSU (available at Digi-Key) producing 12 VDC. This voltage is further reduced to 5 VDC using a linear regulator 7805 outfitted with a heat sink. A current transducer type CAS 6 from LEM is a magnetic transconductance amplifier based on flux core saturation that converts current in its primary winding into the output voltage. This voltage is fed into channel 2 of the ADC. The transducer provides 3 turns of its primary winding that can be connected in series or in parallel to adjust sensitivity via ampère-turns. In our case, the turns are connected in series for maximum sensitivity. Note, the current transducer works with both DC and AC currents.

The JAE connector of the Pi-logger provides stabilized 12 V power and two inputs, digital and analog. The digital input goes directly to GPIO 4 of the Raspberry Pi bypassing the ADC hat, while the analog input is connected to channel 3 of the ADC. This arrangement allows measuring solar or wind power generation characteristics using the pyranometer or a wind sensor, respectively. The pyranometer (discussed below) requires 12 V power and produces digital output over the 1-wire interface. A wind sensor usually also requires 12 V power and produces analog voltage output proportional to the wind speed. The two BNC connectors provide input for voltage measurements in the interval of –10 to +10 V. The black (-) and red (+) spring-loaded contacts provide input for the current measurements.

A rather simple attenuator circuit was added to the Pi-logger to expand the range of measured voltages, Figure 1 inset. The attenuator's input resistance is rather low, about 100 k-ohms, which makes the attenuator not an "ideal" instrument. Nevertheless, for our purposes of measuring battery voltages, the attenuator is perfectly adequate.

Calibration

Although the datasheet of CAS 6 current sensor provides a response curve, it is important to calibrate current measurements. We wrote a small program **voltread** (see Appendix) that prompts the operator to input the known calibration value followed by reporting measured voltage averaged over several measurements. The source code is shown in the Appendix. The code is to be copied into a voltread.c file and compiled with gcc directly on the Pi-logger. The reader is encouraged to inspect the code to understand how to communicate with MCC 118. For CAS6 calibration, the program was invoked as follows, where the parameter -c *is a bit mask selecting channel 2, and the parameter -n indicates averaging over 10 measurements:*

```
sudo ./voltread -n 10 -c 4 > calib_current.txt
```

The attenuator was calibrated using the same program. The calibration results of the current transducer as well as the voltage attenuator are shown on Figure 2. Note the exceptional linearity and reproducibility of the measurements. At zero current, the transducer produces about 2.5 V, depending on the direction of the current, the voltage will either increase or decrease from that value.

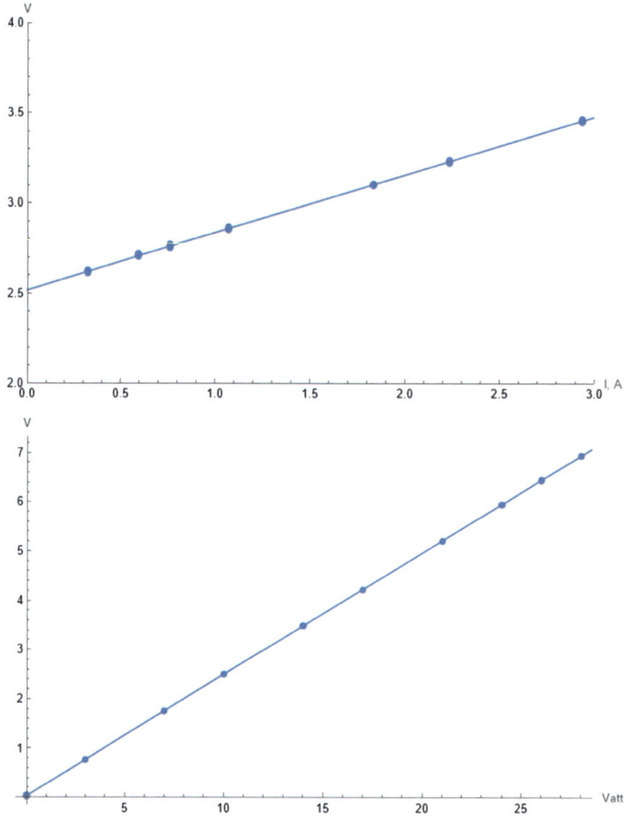

*Figure 2: Calibration of the Pi-logger. Top, CAS6 calibration, V = 2.51641 + 0.320009*I; bottom, attenuator calibration, V = 0.0289635 + 0.246728*Vatt.*

Pyranometer

Realistic expectations from our solar power plant (described below) will depend on how much solar radiation is delivered to our location. The correct instrument for the job is called a "pyranometer", which measures global irradiance. Global irradiance is the amount of energy per unit of time and unit of surface area (*i.e.*, flux, W/m^2) coming from a hemisphere centered at the axis of the pyranometer. This type of irradiance includes direct and diffuse sunlight. In our case the surface of interest is tilted, thus the hemispherical field of view is essentially the sky dome!

A classical pyranometer is made of a black disk enclosed in a wind-proof transparent enclosure, Figure 3. As the disk absorbs the photons, its temperature increases. This heat is transferred into a heat sink (the body of the pyranometer) resulting in establishing an equilibrium temperature that is higher than that of the heat sink. The equilibrium temperature is proportional to the incoming energy flux. Another identical disk is placed in the dark such that its temperature is equal to the temperature of the heat sink. The temperatures are measured by a collection of thermocouples, which provide a voltage output, which is proportional to the insolation [17].

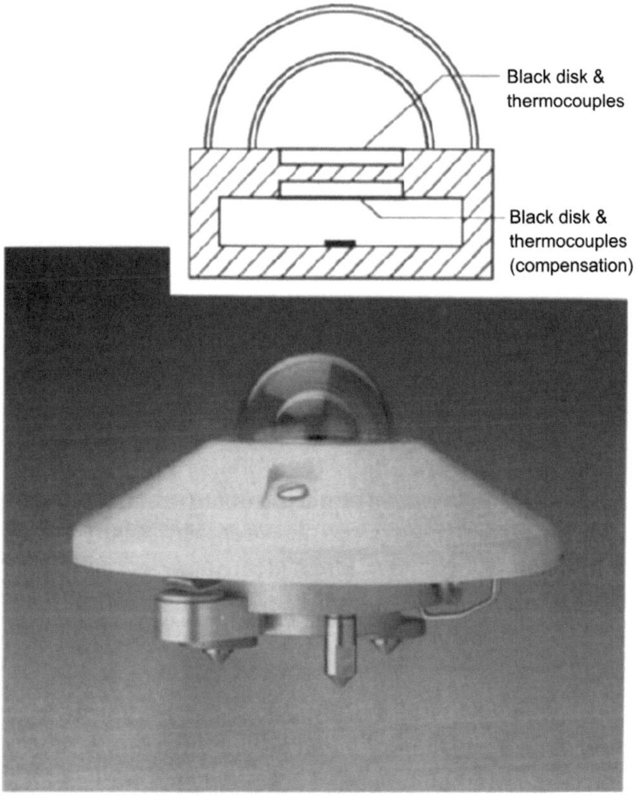

Figure 3. Classical pyranometer, adopted from ref. 17. A black disk absorbs radiation, heats up, and reaches a dynamic equilibrium with the heat sink. The thermocouples produce voltage as an output. An identical disk is placed underneath in the dark for compensation to account for the heat sink temperature.

There are many electronic pyranometers available that use a photodiode and a calibration function to convert a photocurrent into the insolation. For us, however, such an approach is not very interesting, because we want to measure radiation from the first principles. Let's build our own DIY pyranometer for the sake of experience with measurements and understanding the underlying physics behind energy emission/absorption. Also, the DIY pyranometer will be rather inexpensive! Our pyranometer will be like the classic one but with slightly different principles. A schematic of the pyranometer is shown in Figure 4.

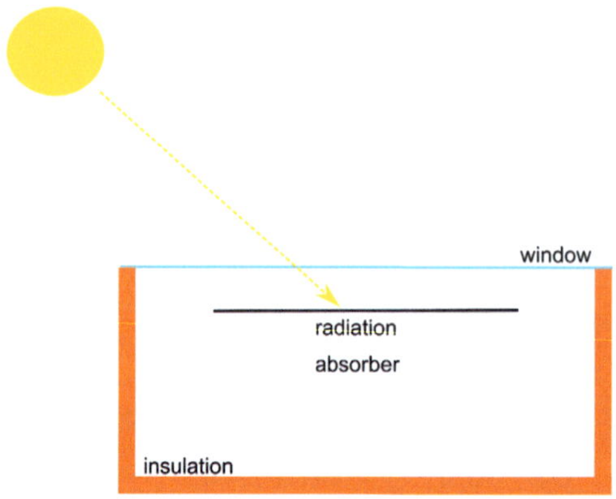

Figure 4: Schematic representation of the DIY pyranometer.

In the DIY pyranometer, the solar radiation flows through a window onto a black radiation absorber. As a result, the temperature of the absorber increases. The temperature will not increase infinitely, because the absorber radiates energy back through the window into Space. This process is governed by the Stefan-Boltzmann law. Specifically, at certain temperature, the amount of absorbed radiation will be equal to the amount of energy radiated back, hence the temperature will become constant:

$$wA-\sigma\left(T(t)^4-T_a^{\,4}\right)\left(A_1\varepsilon_1+A_2\varepsilon_2+\ldots\right)=\frac{dT(t)}{dt}Ah\rho C_p \qquad \text{Eq. 1}$$

Where

- w is an incoming energy flux, W/m2;
- ε_i is the emissivity coefficient of a particular surface, between 0 and 1;
- σ is the Stefan-Boltzmann constant, 5.6703×10-8 W/(m2K4);
- *T* and *Ta* are absolute temperatures of the absorber and the ambient environment respectively;
- A is the area of the incoming energy;
- Ai are areas of the outgoing energy. The emissivity coefficient is a property of a material.

For example, ε =1 for a black body and near zero (e =0.025) for polished gold [18]. It is also important to note that good emitters are good absorbers and vice-versa. Hence, we will not be using polished gold for the absorber, but rather a black painted sheet of aluminum.

The actual design of the DIY pyranometer is shown in Figure 5. The heat absorbing element — a sheet of metal — is situated on two thermally insulating rods, which are placed into the absorber shell — a black cylinder. The absorber shell and the window form a chamber in which the radiation absorber only interacts with the environment via radiation but not thermal conduction. The absorber shell is placed into a plywood box and it is subjected to forced airflow to maintain ambient temperature (air). Therefore, the heat absorbing element re-emits the absorbed energy through the window as well as into the bottom of the absorber shell, whose temperature is maintained constant. The window is made of a ¼ in. thick SUVT Plexiglas, which is transparent to a wide range of light spectrum (see Appendix for datasheet).

Our instrument needs calibration, so a known source of energy is necessary. Since the radiation absorber converts photons into its internal energy and radiates it back, it does not matter how energy is supplied. The simplest and the most accurate way is to electrically warm up the absorber with a tape heater on the back side of the absorber. The electrical power applied to the heater will produce the calibration flux $w = I \times V/A$, where I – current; V voltage; A area of the absorber.

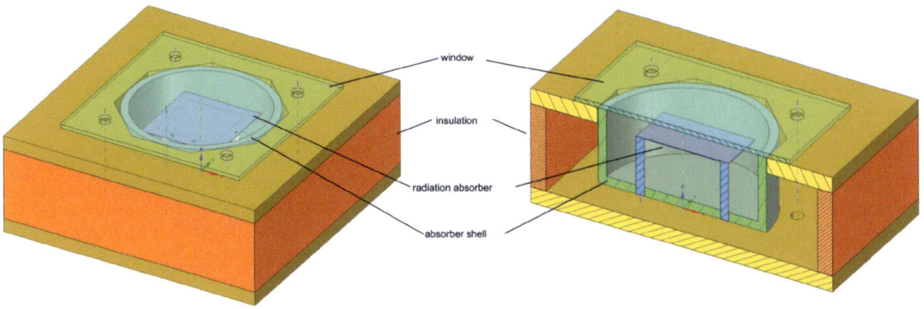

Figure 5. A sketch of the DIY pyranometer (left) and its cross section (right). For calibration, a film heater is placed on the underside of the radiation absorber. See CAD file pyranometer.rsdoc for dimensions and details.

An interesting question is what size shall we choose for the radiation absorber? First, as much as possible we want to cover the inner space of the absorber shell, which can be a 4" sewer pipe cap. A reasonably good dimension is 2.75 × 2.75". What about the thickness? Obviously, the sheet must be thick enough to sustain its own weight as well as a temperature sensor attached to it. The upper limit of thickness is determined by the desired response rate. Indeed, the time course of the temperature as the absorber approaches the equilibrium is as follows:

In this equation, the first term is the incoming energy onto the area A of the absorber; as mentioned above, the second term is the radiation loss from various surfaces of the absorber, which may have different areas and emissivity coefficients (A_i, ε_i), e.g., one side is black painted while the other side is bare metal. The rightmost term shows the rate of change of the temperature multiplied by the heat capacity C_p and the total mass of the absorber, $Ah\rho$, i.e., the product of area, thickness, and density, respectively. For the initial modeling, it is safe to assume that only two major areas of equal size A (top and bottom) are emitting the radiation. Hence, the A may be cancelled; the final differential equation is as follows:

$$w-\sigma\left(T(t)^4-T_a^{\ 4}\right)\left(\varepsilon_1+\varepsilon_2\right)=\frac{dT(t)}{dt}h\rho C_p \qquad \text{Eq. 3}$$

his equation does not have a quadrature solution; therefore, we will solve it numerically using Wolfram Mathematica (Wolfram Research Inc). The desktop version of the software is rather expensive, but a cloud account is free. In Mathematica, the computational workspace is called a "notebook". As mentioned at the beginning of the book, the reader will have to become an engineer to a certain extent; hence learning technical computing products like Mathematica is highly recommended. One important characteristic of Mathematica is its "World knowledge". For instance, density and specific heat of aluminum can be directly obtained within the notebook by a function call. Also, Mathematica is capable of computing with physical units, which is very helpful to making sure the results make physical sense. The numeric solutions for w in the range of 250-1000 W/m2 are shown in Figure 6. The parameters used in the calculation are as follows: Ta = 293 K; e1 = 1; e2 = 0.9; h = 0.01"; Cp = 904 J/(kg.K); r = 2700 kg/m3. The notebook pyranometer_model.nb is provided as a supplement.

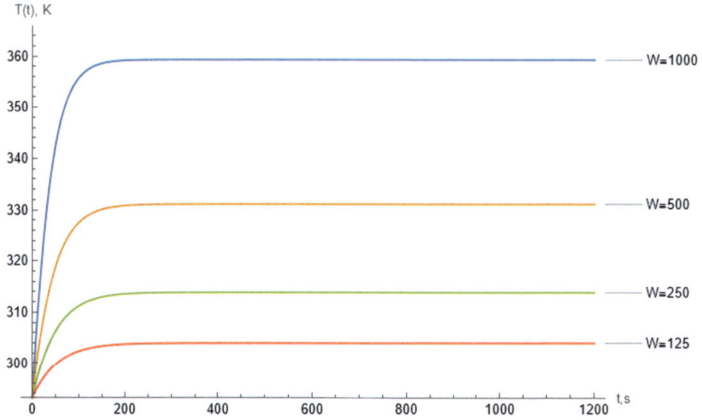

Figure 6: Numerical estimation of the time course of temperature of the radiation absorber at 1000 W/m^2. See text for details.

The time course curve suggests that with thickness of 0.01", the equilibrium will be reached within approx. 200 s, which is very quick and acceptable for our purposes.

What about the heater power? Given the desired calibration energy flux of approx. 1000 W/m^2, an easily available film heater of 2 × 2" (2.581.10^{-3} m^2) would have to be at least 2.6 W.

Pyranometer construction

Dimensions and the detailed 3D model of the pyranometer are provided in the DesignSpark Mechanical CAD file pyranometer.rsdoc. We begin the construction process from the radiation absorber, Figure 7 shows all the steps involved. First, cut a 2.75 × 2.75 × 0.01" sheet of aluminum. Second, make a heat conductive clamp (brass) to hold the digital temperature transducer DS18B20 (purchased from Mouser; datasheet in *Appendix*). Unlike the traditional pyranometer that uses thermopiles and voltage output, the digital output has two main advantages. First, given that the temperature digitizing is taking place within the DS18B20 device itself, we do not need to worry about the noise in the transmitting cables. Second, there is a plethora of thermometers compatible with DS18B20, which features "1-wire" interface including our Pi-logger.

The clamp can be made from a 0.25" wide strip of 0.016" thick brass (available at hardware / hobby stores) that is pressed into a wooden die to form a u-shaped profile (Figure 7A). The die is a block of wood of reasonable dimensions, e.g., 2 × 1 × 1", having a hole drilled in its center with a drill #9 (0.196") and cut in half across the hole. Use a shank of the drill #20 (0.161") as a punch and apply pressure with a vise. By the way, such clamp may be handy to other devices in TO-92 package. Attach the temperature transducer with the clamp and a pair of #4 screws and nuts. On the other side of the radiation absorber, place a 20 W poliimide adhesive heater (Kapton), size 2 × 2". Finally, paint the radiation absorbing side with black spray paint and glue the absorber onto the two 0.25" diameter polystyrene rods held in the absorber shell by friction. The absorber shell is a 4" sewer pipe end cap. In the middle of the cap, place another DS18B20 to measure the shell's temperature. The cap, rods, aluminum sheet and the heater may be obtained from McMaster-Carr.

Figure 7: Steps for making the radiation absorber. A: sheet of aluminum, temperature transducer DS18B20, brass mounting clamp; B: radiation absorber with temperature transducer attached; C: Kapton calibration heater adhesively attached on the side opposite to the thermometer; D: black painted radiation absorber and the shell into which the absorber is to be mounted on the polystyrene posts.

Finally, the absorber-shell assembly is inserted into a box (plywood or plastic) covered with a back wall outfitted with a fan and two "exhaust" copper pipes. The box should be painted white to minimize absorption of solar radiation. The window is made of a SUVT acrylic (Figure 8). Route all the wiring to the outside using tie posts and / or a pin connector, *e.g.*, JAE, yellow and grey respectively in Figure 8.

Figure 8: Final assembly step of the pyranometer. Note the copper pipes for venting air, which are to be inserted inwards of the pyranometer box.

The electrical arrangement of the pyranometer components is shown on Figure 9. Both thermometers, U2 and U3, share the same 1-Wire bus.

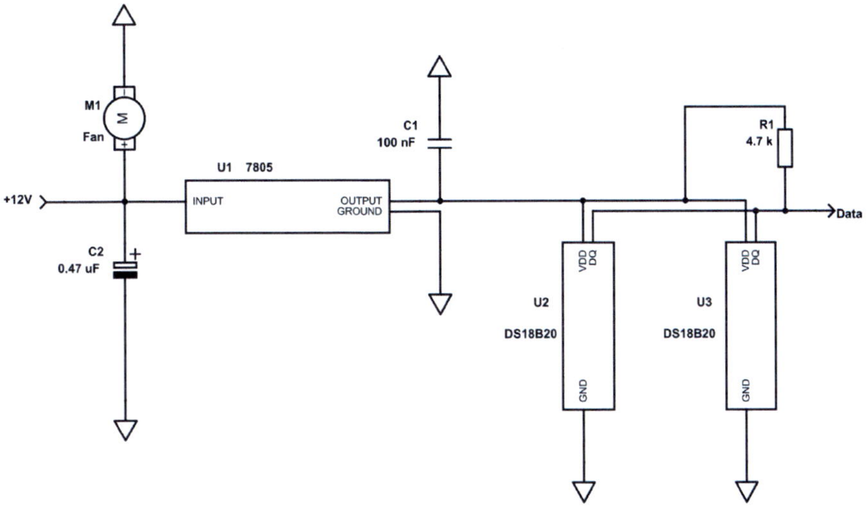

Figure 9: Inside the pyranometer: a fan is directly powered with 12 V; thermometers are powered at 5 V. "Data" is the 1-Wire bus.

Calibration

After the pyranometer is assembled, it is time to calibrate it and assess if our physical model is correct. In fact, we will do two calibrations: one is to calibrate the heater and two is to calibrate the measured temperature versus applied known heat flux. The importance of calibrating the heater is to be able to use just the voltage as a measure of provided energy, rather than measuring both voltage and current. We record characteristics of the heater, which is powered from a variable transformer as follows, Table 1.

VAC	I [mA]	U2 [V]	W/m2
5.1	7.16	26.01	7.48
10.02	14.26	100.40	29.29
16.6	23.1	275.56	78.59
...	...	...	...

Table 1: Voltage, current and heat flux of the 4 in^2 film heater.

As well-known from the Joule–Lenz law, heat output is directly proportional to the square of applied voltage. Graphically we see a perfect straight line, as expected.

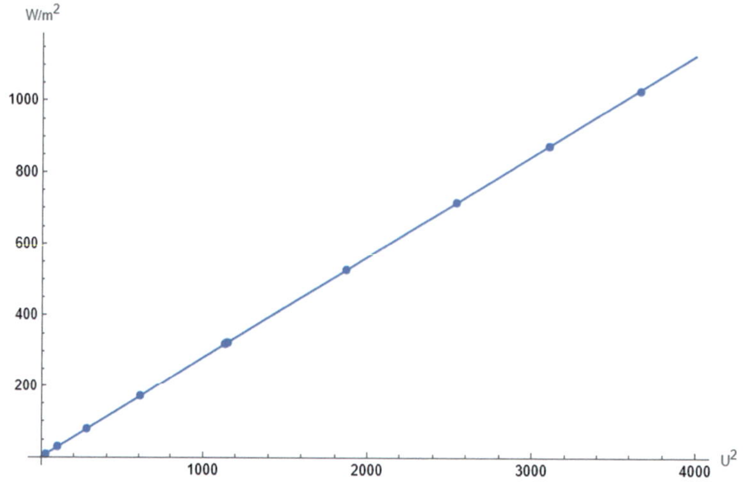

*Figure 10: Calibration of the heater, w= 0.2818*U2; R2=1.*

Knowing the heater characteristics, we can now finally calibrate the pyranometer. We will simultaneously measure the temperatures of the radiation absorber and the temperature of the shell. Using the Pi-logger it is easy to measure temperature of both thermometers simultaneously rather frequently.

As we mentioned earlier, the reader will have to become an engineer. Therefore, learning Linux shell and setting up Raspberry Pi would be very helpful. To enable Raspberry Pi to read the DS18B20 thermometers, the data line (Figure 9) must be connected to GPIO 4 and pulled up to +5 V with a 4.7 k-ohm resistor. Grounds of the pyranometer and Pi are connected as well. Config.txt is to be modified by adding the following line:

```
dtoverlay=w1-gpio;
```

It is important to make sure there are not more than 80 lines in the that file! Restart. Issue the following commands:

```
sudo modprobe w1-gpio
sudo modprobe w1-therm
```

Shut down. Connect the pyranometer. Turn the Pi-logger on. Notice in the directory /sys/bus/w1/devices/ a subdirectory looking like 28-0000000aabc2. This is the internal ID of the thermometer. Since the pyranometer has two thermometers, two of such subdirectories will appear. For all subsequent restarts of the Pi-logger, there is no need for more modprobe commands. The following code makes 230 measurements about every 5 seconds:

```
for i in {1..230}; do sleep 3; echo -e `date +%s`'\t4\t'`cat /sys/bus/w1/
devices/28-0000000aabc2/temperature`'\t'`cat /sys/bus/w1/devices/28-0000000cb89a/
temperature`; done | tee -a templog.tsv
```

The data will be stored in templog.tsv and will be visible at the standard output (thanks to the tee command). The records will look like this:

```
1664946557      4      31437   24500
1664946562      4      31687   24562
1664946567      4      32000   24562
1664946572      4      32187   24625
```

The first number is the number of seconds since 1970-01-01 00:00:00 UTC (quite a big number!); the second is the trial number, e.g., 4; the third is the temperature of the energy absorber in °C multiplied by 1000; the fourth is the temperature of the shell.

Let us apply several voltages to the internal heater of the pyranometer to investigate the heating and cooling characteristics of the pyranometer. A critical characteristic of our instrument is the response time, which we estimated earlier (Eq. 3, Figure 6). Actual response time is shown on Figure 11.

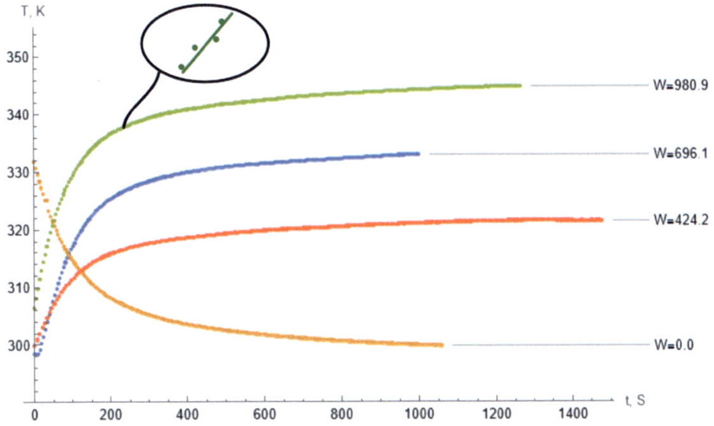

Figure 11: Dynamic characteristics of the pyranometer at various heating regimes W (W/m²). Note, at W=0.0, the curve indicates cooling. An inset instructs measuring the slope of the curve at a particular time point.

As one can see, after about 200 s, the temperature increase begins significantly slowing down. But unlike our model, the transition is not as sharp as calculated. This is the difference between theory and reality. It is not very practical to wait for the equilibrium to be reached, especially in the field, because the position of the Sun is changing throughout the day and clouds may obscure the sunshine for some time.

Instead of waiting for the equilibrium, we can use the dynamic behavior to instantaneously measure the energy flux by rewriting Eq. 3 as follows (Eq. 4):

$$\frac{dT(t)}{dt} = aw - b\left(T(t)^4 - T(t)_a{}^4\right) \qquad \text{Eq. 4}$$

Where *a* and *b* are fitting parameters. The temperature derivative at a certain time *t* is easy to measure by fitting a straight line through 4 points forward; the slope of this line is the derivative (Figure 11 inset). is measured directly. Fitting for *a* and *b* is done using the least squares method. Once the parameters are determined, the energy flux is calculated by solving Eq. 4 for *w*. In our calibration experiments we took about 20 equally spaced time points from three temperature curves (Figure 11) corresponding to fluxes 696.1, 0.0, 424.2 W/m2, performed fitting, calculated *a* and *b*. See *Appendix* for the Mathematica notebook pyranometer_calib.nb and templog.tsv. Based on the fitted parameters, $a=0.000427035$, $b=8.55887*10^{-11}$, we back calculated *w* in all time points and for all fluxes, including 424.2 W/m^2. This test resulted in excellent agreement between calculated and expected energy fluxes, Figure 12, even in the case of 424.2 W/m^2, which was not used in "training" of the model. This result also indicates that our model accurately captures the physics of the process.

Thus, the calibrated pyranometer may be used to record the irradiance as a function of time during the day, in different seasons and at various cloud coverage.

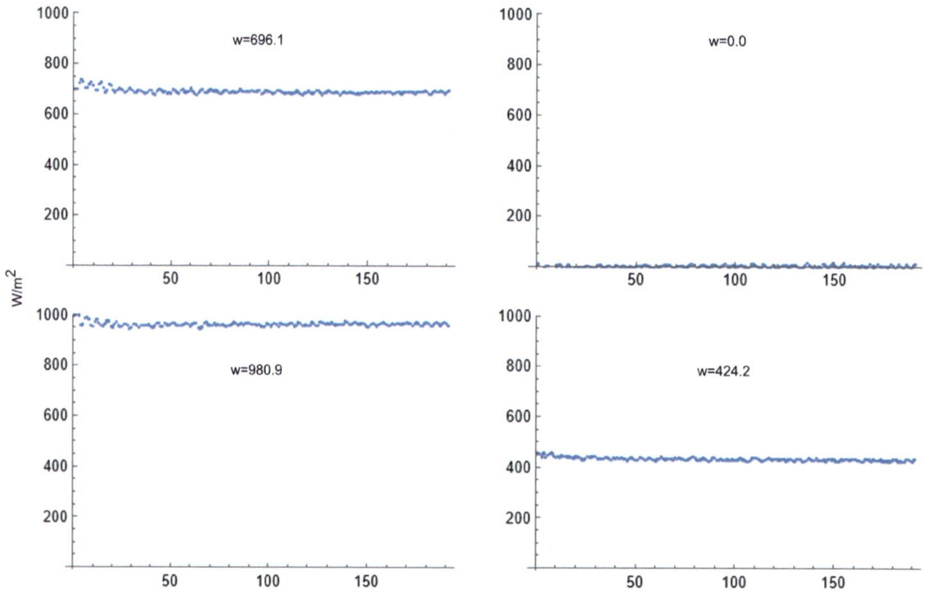

Figure 12: Back calculation of the energy flux w based on parameters $a=0.000427035$, $b=8.55887*10^{-11}$ and measured T(t), T(t)a, and dT(t)/dt .

Chapter 2 • Power from the Sky

Theoretical Foundation

The purpose of this section is to provide a springboard for understanding the processes taking place in PV. The physical model of the PV foundation is quantum mechanics, which formally describes the processes and provides necessary calculations. Interested DIYers are advised to take a course of quantum mechanics, either at school or from multiple free resources, e.g., MIT Open Courseware that was available at the time of writing. Our garage power plant will not require deep understanding of quantum mechanics though. Also, PV Education resource (https://www.pveducation.org/) provides a detailed application of quantum mechanics and solid-state physics to PV.

Photovoltaic effect

This effect was first discovered by a French scientist E. Becquerel, in 1839 who first detected a generation of photo-current in an electrochemical cell (11). The cell was made of an acidic solution, 2 platinum plates and a membrane separating the cathodic and anodic zones. Illumination of one of the plates produced an electrical current. Becquerel also noticed an effect of the light color on the strength of the generated current.

Historically, the findings of the PV effect led to another one, the **photoelectric** effect, discovered by H. Hertz during his experiments in radio wave transmission, further investigated by W. Hallwachs and A. Stoletov. Specifically, Hallwachs (15) charged a zinc plate attached to an electroscope having golden foil leaves as charge indicators. Upon illumination of the plate with an arc lamp, magnesium flame or a stearin candle, different effects were observed. The electroscope connected to the positively charged plate showed very slow discharge (several minutes) when illuminated with an arc lamp or magnesium. To the contrary, negative charge quickly (seconds) disappeared when illuminated with the same light sources. Candlelight failed to produce any charge dissipation of either positive or negatively charged plates. Hallwachs concluded that it was mostly the ultraviolet light that caused the discharge. A theoretical explanation of the photoelectric effect was given by A. Einstein. The photoelectric effect revealed a fundamental property of photons that the energy of photons is dependent on their frequency, rather than the intensity of light. The equation reads $E = hv$, where E is energy, h – Planck's constant and v – frequency. This leads to the following: you may have a very intense light source, but if the frequency of photons is not high enough, no photoelectric effect will be observed.

Bandgap

The photoelectric effect brings us to the basic principle behind the light-induced electricity generation by a solar panel, which is essentially a large collection of connected photodiodes. The energy of photons is absorbed by the electrons that brings them to another energy level allowing them to cross over the **bandgap**. The bandgap is the energy difference of the electron states corresponding to conducting and non-conducting electrons, respectively. The conducting electrons form a so called "electron gas", while the non-conducting electrons are tightly bound to the nuclei. In metals, the bandgap is very small compared to insulators, while the bandgap in semiconductors is in between. Various PV materials have characteristic bandgaps.

The importance of the bandgap for the PV materials is that it directly determines the efficiency of solar energy conversion. Efficiency of light conversion is defined as a power ratio W_{out}/W_{in} produced electrical energy over the incoming energy of photons per unit of time. Figure 13 shows a theoretical calculation of a single-junction silicon-based PV element. One of the curves on the left panel has large dips; this curve corresponds to the solar spectrum at sea level. (More recent measurements of the solar spectrum are provided in Gueymard et al., 14). The spectrum is the photon flux intensity i.e., the number of photons per unit cm^2 per sec., as a function of the photon energy. The yellow shaded area is the total number of photons with energies above the bandgap (e.g., 1.35 eV for silicon). Now let us switch to the right panel, whereby the Y coordinate of each point on the "100%" curve represents the area of the yellow shading as a function of a bandgap energy value. The blue shaded area under that curve is the total energy delivered by the photons with energies above the bandgap. Unfortunately, there is an intrinsic loss due to radiative recombination. The PV element not only absorbs, but also emits light. The rate of the radiative emission exponentially increases with the voltage developed across the load (determined by the laws of thermodynamics and statistical mechanics). The emission results in the equivalent of "radiative current", which subtracts from the current through load. At the maximum power point (MPP, discussed below), the effective curve (formerly 100%) shifts to the left, which is labeled "W" in Figure 13, right panel.

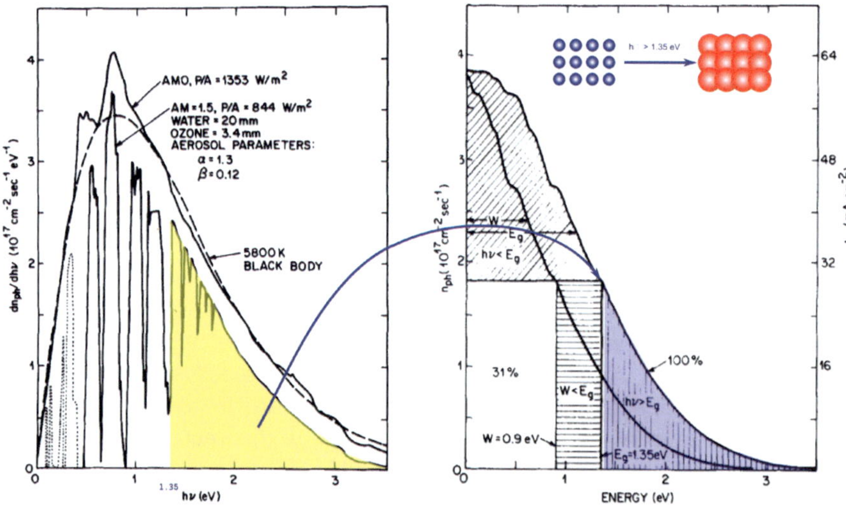

Figure 13: Efficiency of a single-junction PV element with a bandgap Eg=1.35 eV. Left panel, curve with dips: solar spectrum in the Northern Hemisphere, sea level. Right panel: Y coordinate of each point of the "100%" curve is the area under the solar spectrum curve (yellow shading) as a function of a given bandgap energy. Blue shading indicates total solar energy flux with photons having energies above the Eg value. See text for details. Adopted from Henry 1980 (12).

In addition, there are two other losses:

(i) the energy of photons above the bandgap is lost into heat;

(ii) the photons with energy below the bandgap are not absorbed.

All is left is the white rectangle corresponding to 31% of the solar energy that can be converted into electricity. It is important to understand for the material with some other value of the bandgap, the white shaded rectangle will be different. If several junctions are sandwiched together, efficiency may be drastically increased, however such PV panels do not seem to be commercially available at present. Moreover, such cells would require a concentrator (lens or mirror), which is not practically feasible at home. A recent review in the Science magazine provides a survey of existing PV materials, bandgaps and efficiencies (13).

Solar panels

For our garage power plant, we chose Renogy RNG-100D-SS monocrystalline 100-watt solar panels. It is important to understand that 100 watts are produced in standard test conditions, *i.e.,*

- MPP
- irradiance 1000W/m^2
- temperature 25 °C
- AM 1.5

MPP is a condition on the *I-V* curve, at which the solar panel delivers maximum power to the load. Specifically, MPP for Renogy RNG-100D-SS is shown in Figure 14. What is AM?

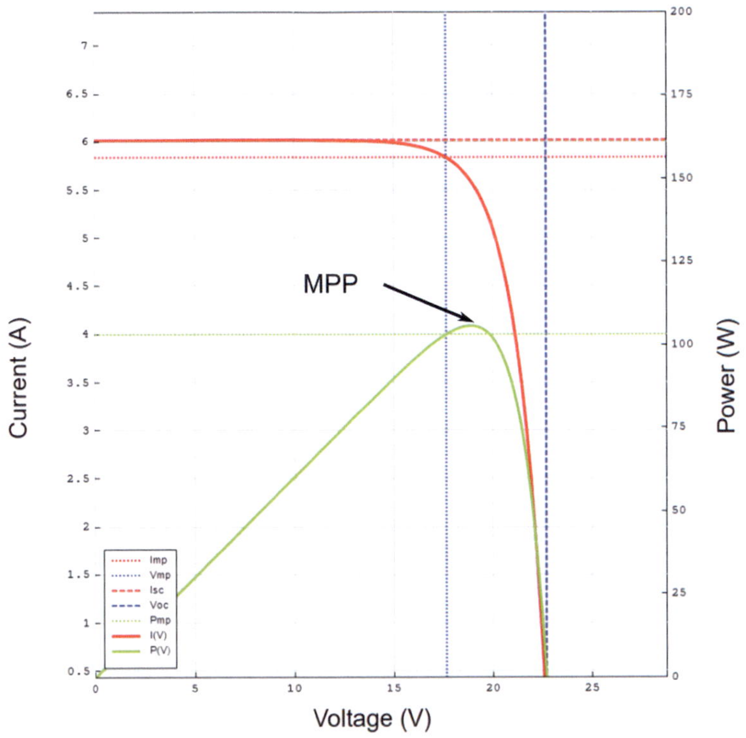

Figure 14: MPP corresponds to the maximum on the power curve (green). Adopted from Renogy RNG-100D-SS datasheet.

RENOGY

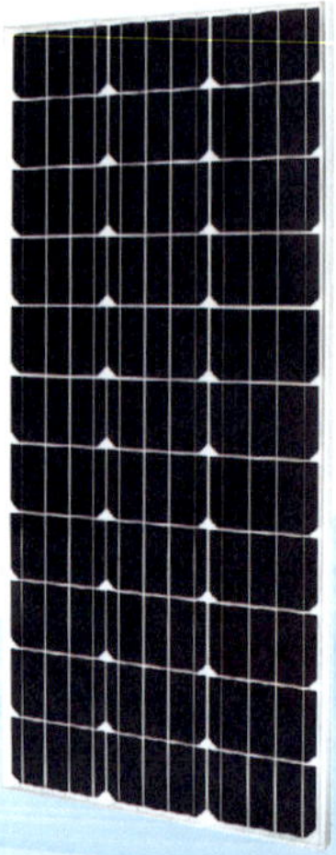

RNG-100D-SS

100W Monocrystalline Solar Panel

Key Features

Sleek design and a durable frame, the Renogy 100 Watt 12 Volt Monocrystalline Panel provides you with the highest efficiency per area and is the perfect item for off-grid applications.

- High module conversion efficiency
- Top ranked PTC rating
- Quick and inexpensive mounting
- 100% EL testing on all Renogy modules
- No hot spots guaranteed

Potential Uses

The Renogy 100 Watt Monocrystalline Panel can be used in various off-grid applications that include 12 and 24 volts arrays, water pumping systems, signaling systems and other off-grid applications.

25 Years — Power Output Warranty

5 Years — Material and Workmanship Warranty

Renogy | www.renogy.com | support@renogy.com | 909-287-7111
2775 E. Philadelphia St., Ontario, CA 91761

RNG-100D-SS

100W Monocrystalline Solar Panel

Electrical Data

Maximum Power at STC*	100 W
Optimum Operating Voltage (V_{mp})	17.9 V
Optimum Operating Current (I_{mp})	5.72 A
Open Circuit Voltage (V_{oc})	21.6 V
Short Circuit Current (I_{sc})	6.24 A
Cell Efficiency	21.0%
Maximum System Voltage	600 VDC UL
Maximum Series Fuse Rating	15 A

Mechanical Data

Solar Cell Type	Monocrystalline (6.1 x 4 in)
Number of Cells	33 (3 x 11)
Dimensions	42.2 x 19.6 x 1.38in (1074 x 498 x 35mm)
Weight	14.3 lbs (6.5 kg)
Front Glass	Tempered Glass 0.13 in (3.2 mm)
Frame	Anodized Aluminium Alloy
Connectors	Solar Connectors
Fire Rating	Class C

Thermal Characteristics

Operating Module Temperature	-40°F to 176°F
Nominal Operating Cell Temerature (NOCT)	47±2°C
Temperature Coefficient of Pmax	-0.23%/°C
Temperature Coefficient of Voc	-0.33%/°C
Temperature Coefficient of Isc	0.05%/°C

Solar Connectors

Rated Current	30A
Maximum Voltage	1000VDC
Maximum AWG Size Range	10 AWG
Temperature Range	-40°F to 194°F
IP Rating	IP 67

Junction Box

IP Rating	IP 65
Diode Type	HY 10SQ050
Number of Diodes	2 Diode(s)
Output Cables	14 AWG (2.00 ft long)

Certifications

Module Diagram

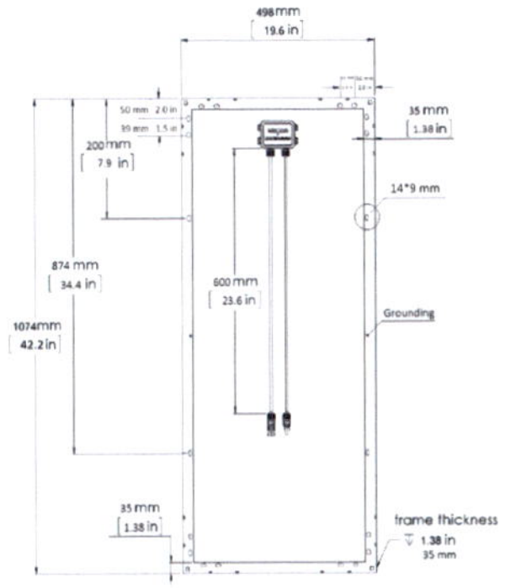

IV-Curve

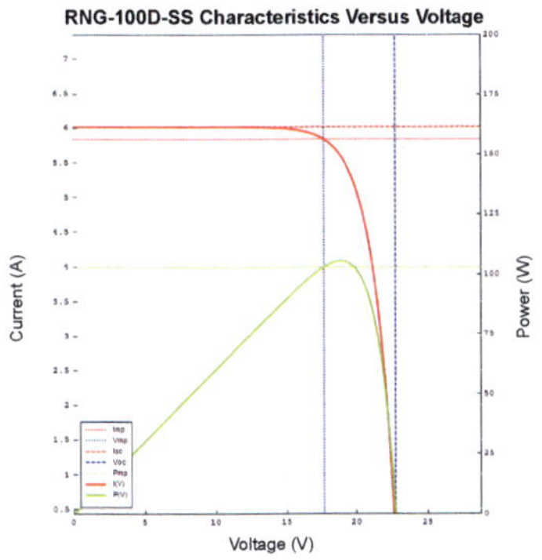

*All specifications and data described in this data sheet are tested under Standard Test Conditions (STC - Irradiance: 1000W/m², Temperature: 25°C, Air Mass: 1.5) and may deviate marginally from actual values. Renogy and any of its affiliates has reserved the right to make any modifications to the information on this data sheet without notice. It is our goal to supply our customers with the most recent information regarding our products. These data sheets can be found in the downloads section of our website, www.renogy.com

Renogy | www.renogy.com | support@renogy.com | 909-287-7111
2775 E. Philadelphia St., Ontario, CA 91761

Renogy RNG-100D-SS datasheet.

Air mass coefficient

The air mass coefficient (AM) indicates how much atmosphere is involved in affecting the solar spectrum. At the average latitude of North America, 37°, the average sun elevation is 41.81° above the horizon. This makes a 48.19° angle between zenith and the light path

(14). A diagram in Figure 15 represents these standardized conditions, resulting in the path length through the atmosphere of 1.5 times longer relative to zenith. The importance of the atmosphere is that part of the light will be scattered. The extent of scattering is dependent on wavelength, the shorter the wavelength the higher the scattering. Remember the dawn and dusk colors? The blue portion of light is scattered as it travels through the long path in the atmosphere, as it occurs at sunrise and sunset. The transmitted light through the thickness of the atmosphere follows the Beer-Lambert's law,

$$I = I_0 e^{-\alpha L}$$

where I is the light intensity (W/m^2) as it is absorbed / scattered in the medium (*e.g.*, air), at the distance L from the point where the intensity is I_0. a is the "extinction coefficient". The extinction coefficient is dependent on the medium and the wavelength of the light.

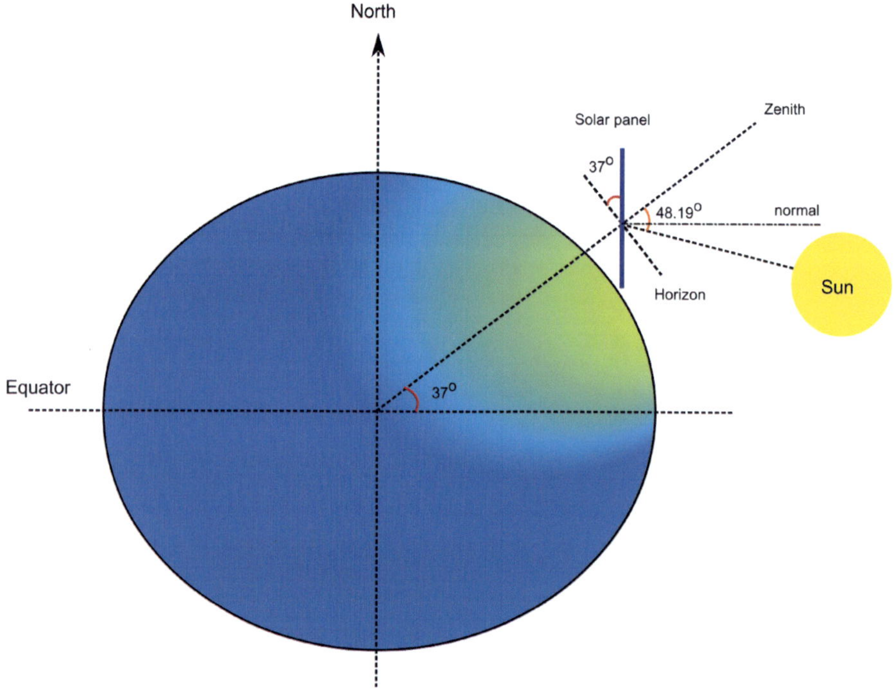

Figure 15: Standardized light path calculation in North America at the average latitude of 37°. Average sun elevation above the horizon is 41.81°, resulting in 48.19° angle between zenith and the light path.

Light management

Interestingly, the same law plays an important role in the way the solar panel absorbs light for conversion into electricity. According to PV Education (https://www.pveducation.org/), the extinction coefficient of the monocrystalline silicon at 550 nm (peak of the solar spectrum at AM 1.5) is 10 times larger than that of the amorphous silicon. Hence, the monocrystalline silicon absorbs more photons than its amorphous counterpart over the same depth from the surface into the material. Also, the surface of the solar panel is roughened to create ridges and valleys enabling multiple light interactions through reflection therein. Finally, the surface is coated with an anti-reflective coating sending the light back to the surface (Figure 16). These approaches are called "light management" that intend to increase *L* — the optical path — in the Beer-Lambert's equation. The visible effect of the light management is that the panel looks dark due to effective absorption of the light.

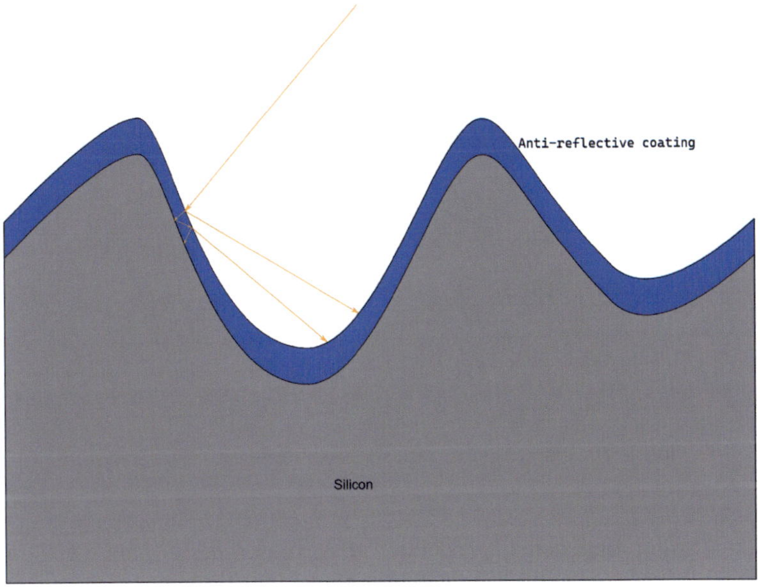

Figure 16. Light management is accomplished by roughing the surface and providing an anti-reflective coating.

Garage Powerplant

As we finished with the theoretical foundation, it is time to create our own powerplant! We will use Renogy RNG-100D-SS monocrystalline 100-W solar panels, which are very light and easy to handle. Renogy provides short (about 4 inch) mounting brackets, which work well on an "ideal" surface — flat and smooth. However, if the roof is covered with tiles, one needs a different type of mounting. We designed a bracket based on a long aluminum z-bar. The long bracket allows locating the best tethering position, *i.e.,* the one with smallest gap, on the uneven tiles (Figure 17).

Figure 17: Solar panels mounted on an uneven surface with z-bar brackets.

Mounting Solar Panels

Every roof is different and particular configurations may require unique solutions. Nevertheless, the mounting presented here will likely work on most of the roofs, regardless of how uneven they are. The mount is a z-bar made of 1/16" thick aluminum, 1-1/8" height, 1-1/8" wide, 3 ft long. The z-bar is available from the metal supplier stores. Holes in the z-bars and panel frame are drilled through a jig, which correctly positions the holes without needing to perform repeated measurements. The jig is made of a steel 1 × 39 × 1/8" bar having 1/8" holes drilled according to the appropriate pattern. CAD files of the jig and the mounting bracket are provided as a supplement.

Figure 18 shows all necessary steps to attach the mounting brackets. First, drill 1/8" in. pilot holes in the panel frame and z-bars. Second, use drill #9 to enlarge the z-bar holes. Attach the bracket to the panel with #8–1/2 galvanized sheet metal screws. The sheet metal screws cut their thread in the 1/8 in. holes of the panel frame.

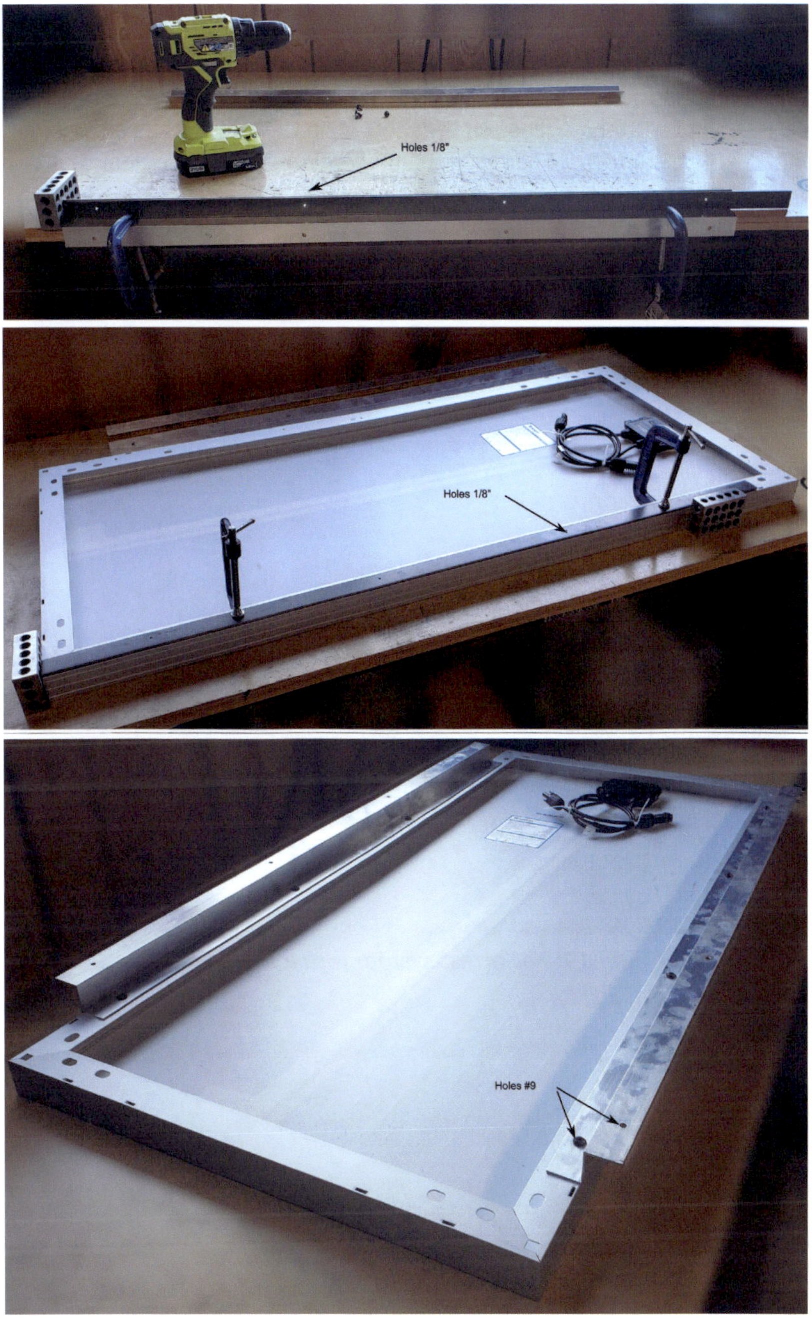

Figure 18: Drilling holes in z-brackets and PV panels with a jig; attaching brackets with sheet metal screws.

On the Roof

While working on the roof, use appropriate fall protection gear. Usually, the fall protection is a thick nylon rope with a carabiner on one end and a lanyard with a sliding grip. The carabiner is attached to a strong anchor on the roof, while the lanyard has a carabiner as well, which connects to the harness. There is also an energy absorbing element that decelerates the body during the fall thus minimizing the whiplash.

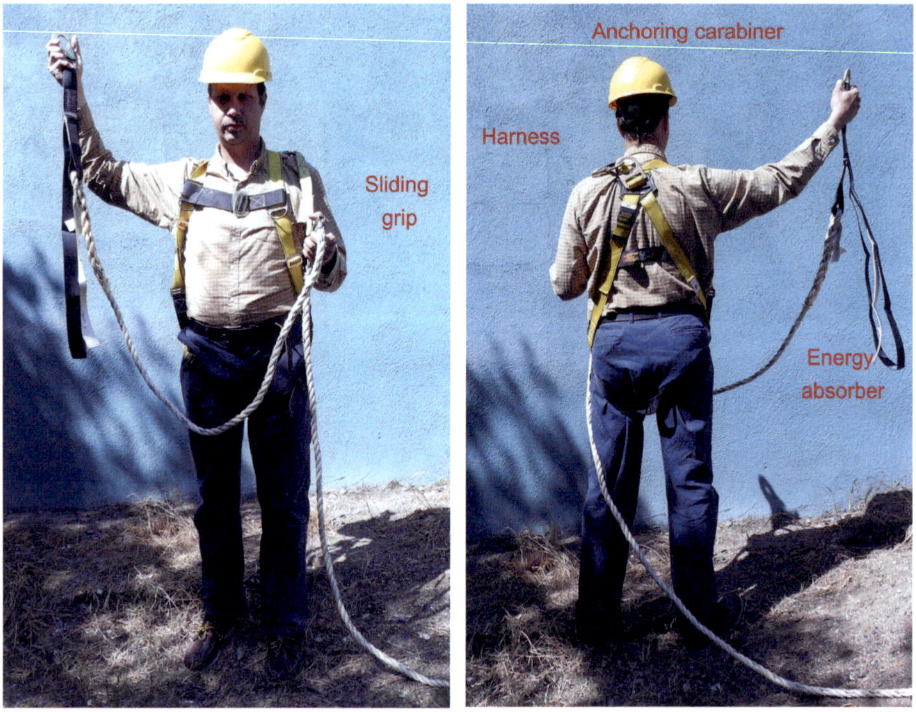

Figure 19: Fall arrest system with a sliding grip.

Place the panel with attached mounting brackets onto the roof in a desired location. Figure 20 illustrates the steps involved. The mounting brackets provide several holes for locating the optimal fastening points (as an example, see red markings on Figure 20). Drill holes with a 5/32 masonry bit and a hammer drill. Fill the holes with a waterproofing material, for example, Gardner-Gibson Black Jack Neoprene Super Flash Cement. Fasten the panels on the roof with screws of sufficient length, providing the screw penetration through all layers of the roof (e.g., #10–3-1/2 Hillman multi-material screw).

Figure 20: Placement of PV panels on the roof. Top to bottom: optimal holes located, marked red; holes drilled, waterproof cement added; panels screwed on. Observe black waterproofing cement around the screws.

From Photons to Electrons

As we have finished installation of the panels, it is time to collect the energy in the form of electrical current. The first question we must ask ourselves is how to connect our PV panels. (A collection of panels is called a "PV array"). Two obvious ways are series and parallel. In the absence of shading at MPP, N panels in series will deliver V = NVMPP and I = IMPP, while the same number of panels in parallel will deliver V = VMPP and I = NIMPP. Resistive losses in the wiring, connections, and FETs inside of the charger increase with current: $P_{loss} = RI^2$, where R is the apparent resistance of the entire circuit. Clearly, increasing current is detrimental to performance. (By the way, this is the reason why the grid power is transmitted over large distances at a very high voltage). One would think that a series connection is the best configuration, which is true only in the absence of shading. With shading, a panel becomes reverse polarized, and a bypass diode passes current through effectively removing the panel from the array (Figure 21).

It turns out that finding an optimal configuration is not as simple as you may think. A review from 2018 (16) covers the issue extensively. The takeaway from this analysis is that a PV array configured as one string (series) will have several maximums on the power-voltage curve, unlike the one depicted on Figure 14. Such curve poses a serious difficulty for MPP tracking because the algorithm will likely get stuck in one of the maximums. Malathy et al. (16) provide references for interesting solutions that use neural networks and heuristics to search for the maximum, which is currently more academic than practical. A parallel configuration of the PV array will have only one maximum, same as on Figure 14, however as discussed above, high current poses a disadvantage. Apparently, the ideal configuration ensures homogenous distribution of the shaded panels across the array, see Figure 21 bottom. In this case, only one peak on the P-V curve is expected, hence the charge controller will find MPP without problems.

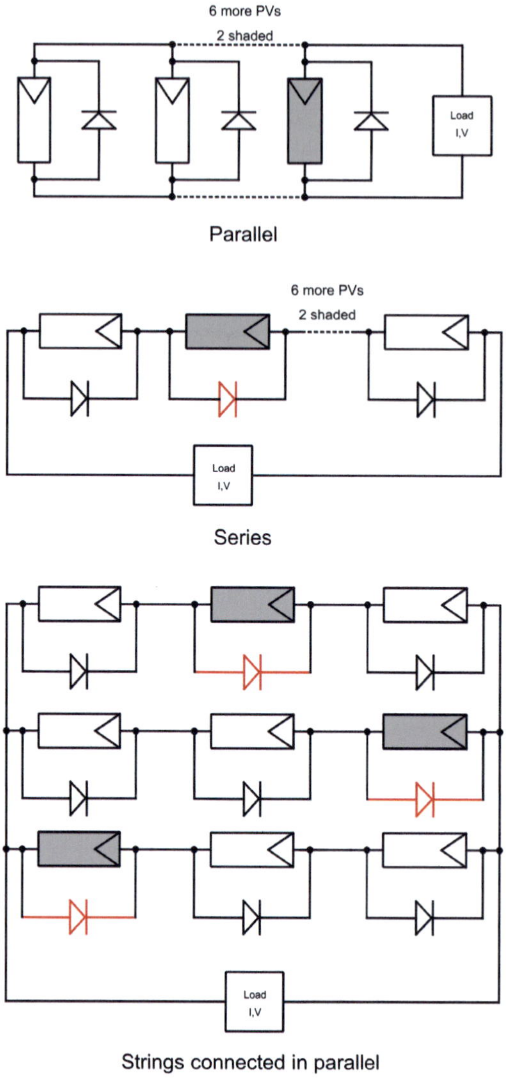

Figure 21: Parallel, series and parallel strings connection configuration of a 9-panel PV array. Grey shading indicates partial shading of the PV array. Bypass diode is connected in parallel with a PV element allowing the current to pass thru in case of partial shading. When the diode is active, it is marked red.

Here is the specific shading pattern of the PV array of our garage powerplant (Figure 22). The study was done on Aug 28-29. On these days in Los Angeles, CA the sunrise was at 6:25, sunset at 19:23. Interestingly, the eastern field of the roof is shaded for almost 5 hours after the sunrise. This is because of a big oak tree on the property that provides a nice shade for the main house and casts some onto the garage. We opted out of the chainsaw solution in favor of tolerating some power loss. The shading pattern suggests an optimal configuration of 3 strings 3 panels each. The east-field panels should be mixed in with west-field panels to ensure that in the morning and in the afternoon each of the 3 strings has just 1 panel shaded. Better yet, one of the panels from the east field may be moved to the west.

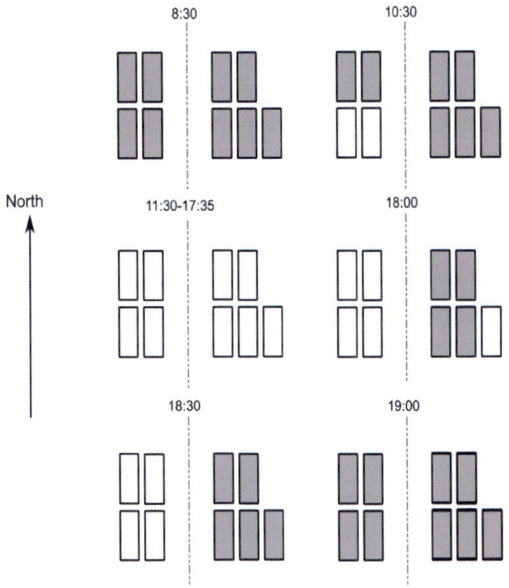

Figure 22: Nine-panel PV array shading pattern on August 28-29. Dashed line indicates roof ridge. Sunrise 6:56, sunset 19:23.

After all connections have been made, rout a cable rated for outdoors, for instance SOOW (oil- and water-resistant insulation, 600 V), through the "service cap" into the building (Figure 23). If all 9 panels work at the MPP, the 3 strings in parallel will produce 12 A. For this current, the cable should carry #14 or #12 AWG wires. As mentioned in the Introduction, refer to the Pocket Ref (1) for determining the ampacity of the wires and insulation codes.

Figure 23: Service cap.

Inside of the building, set up a charge controller, inverter, and a load center with circuit breakers. A schematic of a possible arrangement is shown on Figure 24. S1-S3 are the overcurrent protection circuit breakers. In our case, we established a 5-kW system with the MN5048DIY charger-inverter (see Appendix for datasheet and manual). Importantly, this inverter can take DC and AC input from the grid or a generator. The inverter must also be connected to the grid, such that when the batteries discharge below a certain threshold, the grid would take over to provide power to the garage as well as charging the batteries. Here comes the issue of sizing our power plant: the battery capacity should be sufficient to hold overnight without switching over to the grid. Yet, the battery capacity must be matched with the PV output to be able to charge them during the day.

The circuit breakers S1 and S2 are rated for DC current, 20 A and 125 A, respectively. Note, such breakers could not be found at a regular Home Improvement store at the time of writing. However, there are plenty of inexpensive online options. Also, these breakers will not fit standard house subpanels, therefore a box with a DIN rail is required. In contrast, S3 is a regular AC breaker rated for 50 to 60 A, which should be installed in a separate panel to distribute usable 120 VAC.

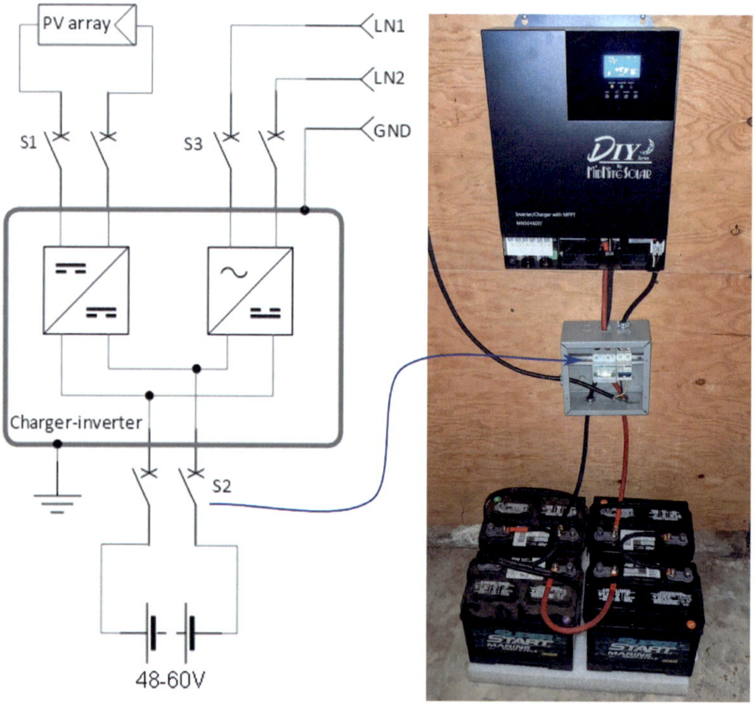

Figure 24: Left: a schematic of an arrangement based on a hybrid charger-inverter, which charges batteries and produces line AC. Right: photo of an actual setup, AC output not yet attached.

The "battery bank" is made of four deep-cycle 12 V marine batteries (24DCMJ), 80 Ah capacity. The wiring for batteries and S2 must be able to withstand very high currents so 2 AWG wire is recommended. Attach the wires to the batteries using lugs that match the wire diameter and the post on the battery. For example, in our case we used the #2 AWG lugs for 5/16" posts. Naturally, you would need a wire cutter, and a crimper to squeeze the lugs around the wires, Figure 25.

Figure 25: Copper lugs and a crimper.

The finalized PV system is shown on Figure 26. The "PV inlet" box is where the PC through the wall is where the PV cables are coming into the building through the service cap and its conduit.

The box where the PV power comes into the building is a perfect place for adjusting the configuration if multiple arrays exist on the roof. For example, as discussed above, some of the arrays may be connected in series, while others are in parallel. The last chapter *Conduits and Batteries* provides a good starting point for understanding and choosing conduits, boxes, and conduit bodies.

Figure 26: Finished photovoltaic system configured to generate and distribute usable power at 110 VAC. Grounding wire (GND) leads to the physical ground outside of the building. Note a small device designated as EQ — an optional battery equalizer.

Let's use the pyranometer to measure insolation near the solar panels. The pyranometer was placed on a large plywood sheet, which was secured on the roof. The plywood provided a flat base for the pyranometer. The Pi-logger was used to collect the data throughout the day using a command that called for a script to run in the background:

```
nohup ./templog.sh &
```

The script templog.sh is as follows:

```
#!/bin/bash
counter=0
while true; do
for i in {1..10000}; do sleep 3; echo -e `date +%s`'\t'`cat /sys/bus/w1/
devices/28-0000000aabc2/temperature`'\t'`cat /sys/bus/w1/devices/28-0000000cb89a/
temperature`; done > templog$counter.tsv
counter=$(( counter + 1 ))
done
```

The obtained temperature measurements were converted into the energy flux according to the pyranometer calibration, Figure 27. Interestingly, for almost the whole morning the panels did not receive any significant sunshine. This is due to the shadow cast by a big tree in our yard, which we are **not** planning to cut!

How much electrical energy, watthours, can we expect to get from our panels given the measured insolation? To answer this question, we need to

1. calculate the area under the curve (Figure 27);
2. multiply by the total solar panel area;
3. multiply by the efficiency of the panels.

Using numerical integration in Mathematica and the solar panel datasheets, we estimate about 5 kWh. This amount of energy should be sufficient for the four 12 V 80 Ah batteries, which require 3.8 kWh. In fact, the charger-inverter consistently reported batteries fully charged in the afternoons of the sunny days in August 2023.

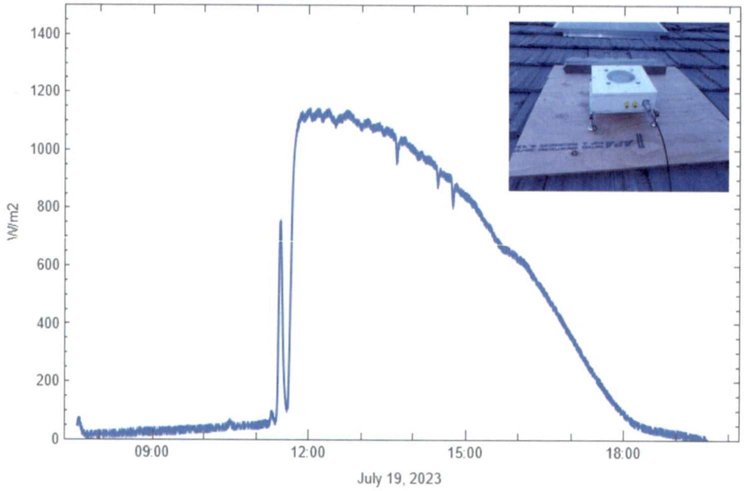

Figure 27: Insolation, measured in Altadena, CA on July 19, 2023. The pyranometer is placed near one of the panels.

Charging regime

Initially we planned our power plant to be 100% off-grid and did not connect the hybrid inverter to the grid at all. After the system was put into production, we were surprised to see the batteries seemingly lose their capacity rather quickly. Without any significant load, after a couple of months, each morning the system sounded an alarm indicating voltage below 44 V and the inverter output at zero. Assessing the battery conditions by measuring electrolyte density and capacity revealed significant degradation. In the last chapter of the book, we discuss how to assess battery capacity and how to revive a lead-acid battery, while it is still possible.

A presumed reason for the battery degradation was allowing batteries to deep discharge and daily charge / discharge cycles. Anecdotal evidence from a friend who is running a similar size off-grid system with a lithium battery set revealed a progressive loss of capacity after about 6 months of production use. The battery was discharged 50% nightly and recharged fully the following day. After 6 months, the battery lost about 25% of its capacity.

To save the batteries, we connected the hybrid inverter to the grid with the intent to switch to the grid when the batteries discharge to a specific level. The following parameters were set on the hybrid charger / inverter:

- Inverter priority mode: the inverter switches to the grid when the battery voltage is below a threshold.
- Voltage threshold for switching to grid, 48.4 V.
- PV charging priority: charging predominantly from the solar. When the system is switched to the grid, then charge from the grid as well.

- Battery undervoltage alarm, 48.4 V. It turned out that the battery / grid threshold and the undervoltage alarm settings were linked together. (The reader's equipment may be different).

- Battery undervoltage recovery point, 52.0 V. As the voltage reaches the recovery point, the system goes off grid.

- Max AC charger current, 5 A.

The last parameter is critical, because at night we do not want to charge the batteries quickly from the grid with high current. If we did allow rapid charge from the grid, in the night the system would switch off the grid and return to using the batteries. After that, the batteries would discharge again making the system switch back to the grid. Such cycling would result in battery wearing out. Note, this parameter only affects AC charging; during the daytime there is no restriction on the PV charging current.

Chapter 3 • Wind Energy

A small "backyard" turbine in a windy area may be a great source of energy, if properly used. Interestingly, a cursory look at the customer comments on Amazon reveals that the turbines receive mixed reviews. The main criticism is that upon actual measurements, a turbine produces insufficient power. Why is this happening? This is because for a given wind speed, the performance of the turbine nonlinearly depends on its RPM, which in turn depends on the load. Let us dive into some theory to understand this "inconvenient" phenomenon so that we could properly harvest energy from the wind.

Power in the Air

The energy delivered by the wind is carried by the moving mass of air, *m* with speed v (Eq. 5).

$$E_{wind}=\frac{mv^2}{2} \qquad \text{Eq. 5}$$

The turbine blades of the lengths *r* cover a disk-shaped area of πr^2, such that the total mass of air coming through this area per second is $\rho\pi r^2 v$, where r is air density. This is because in one second the air extends behind the blades into a "solid cylinder" of length v and base area πr^2. The volume of the cylinder is $\pi r^2 v$. Putting the expressions together, we can obtain the total amount of energy delivered by the wind per second, i.e., power (Eq. 6):

$$P_{wind}=\frac{\rho\pi r^2 v^3}{2} \qquad \text{Eq. 6}$$

We can already see that the bigger the blades, the faster the wind and the denser the air, the more energy is carried. Unfortunately, the turbine cannot extract the total power due to several factors, i.e., aerodynamic efficiency, mechanical loss, and Joule (heat) loss in the windings. The aerodynamic efficiency is quite nontrivial: it originates from lift, drag and wake turbulence. The aerodynamics of the blades determines a fraction of wind power extractable by the blades [19]. This fraction is called a "power coefficient", which depends on the "tip speed ratio" (TSR). TSR is an important concept to understand, as it measures how fast the blade tips are moving with respect to the wind (Eq. 7).

$$TSR=\frac{2\pi rf}{v} \qquad \text{Eq. 7}$$

where *f is* the number of revolutions per second. Interestingly the power coefficient is a curve with a maximum, and it cannot exceed 0.59 (the Betz Limit). Figure 28 shows an example of the power coefficient for some blades. Note, it does not even reach the Betz Limit.

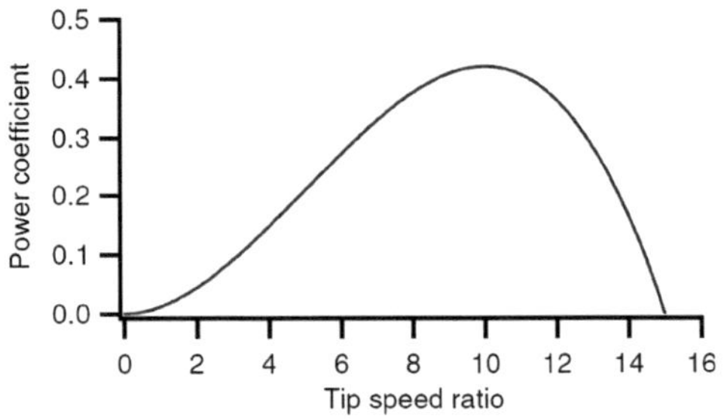

Figure 28: Power coefficient, adopted from Manwell et al [19].

We can appreciate the fact that the extraction of energy from wind is complicated by the interplay of various losses and the power coefficient. Therefore, each turbine needs to be characterized experimentally, *e.g.*, see ref 20. Earlier we have conducted similar experimental research with a Chinook 200 wind turbine (MidNite Solar Inc.). The electrical characteristics of the stator were determined by turning the turbine with a motor on a special stand. The aerodynamic characteristics were measured by driving the turbine on a truck at various speeds to generate a range of wind speeds [21]. In that study, the power coefficient curve was reproduced and, importantly, the voltage of the optimal energy transfer was determined (Figure 29).

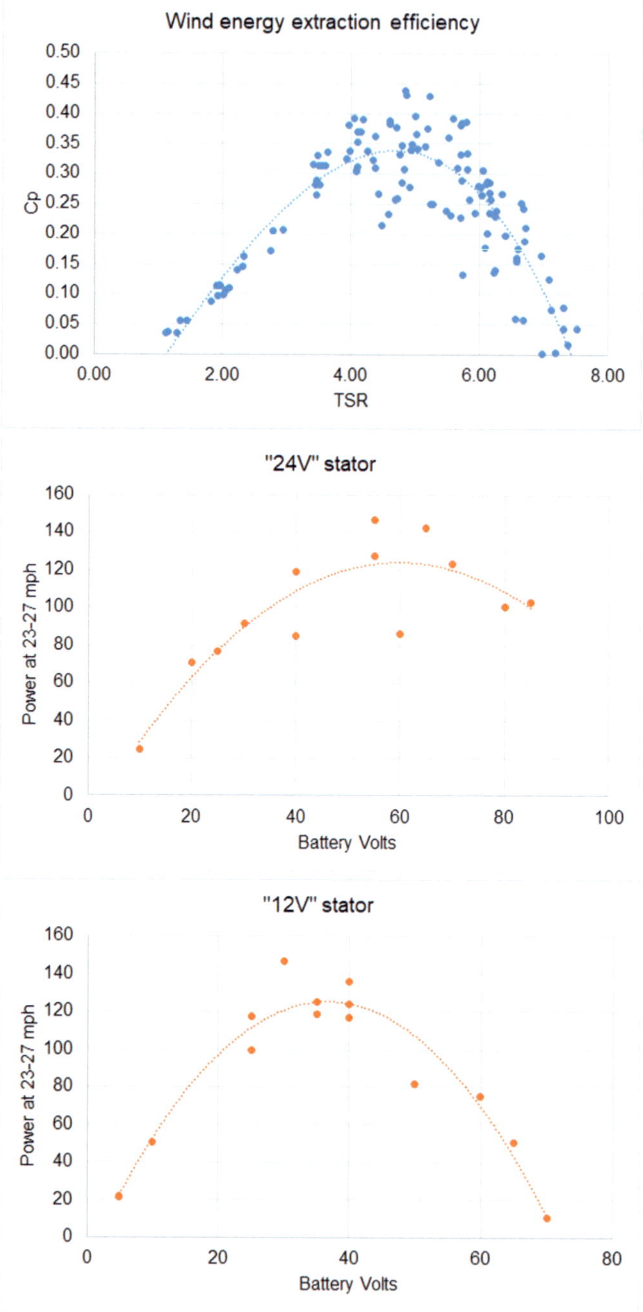

Figure 29: Experimental characterization of the Chinook 200 wind turbine. Top: power coefficient dependence on TSR; middle and bottom — optimization of loading "battery" voltage for the 24 and 12 V stators, respectively. Figures adopted from Ref. 21.

The peak of the power-loading voltage curve is readily understandable. Indeed, at a low voltage, the current is high, the opposing magnetic field is high, and the turbine turns slowly. As it turns slowly, a large portion of air mass passes between the blades without delivering any energy. On the other hand, if the voltage is too high, the turbine must reach high RPM to be able to transfer charge to the batteries. At high RPM, the blades are subjected to a significant drag.

Here we present a practical installation, characterization, and optimization of the Chinook 200 wind turbine (without the need of a special stand or a truck). The same logic and circuit will apply to other types of small turbines.

Turbine Setup

The Chinook 200 turbine characteristics are as follows: rotor diameter 3.28′ (1 m), rated output 200 W at 31.3 mph (14 m/s) wind speed, weight 14 lbs, start-up wind speed 7.8 mph (3.5 m/s). The turbine has a swivel mounting mechanism for a tubular mast. This is a critical part of the turbine that requires attention. Specifically, the generated current is passed through a slip ring mechanism that connects stationary tube and wires with the swiveling turbine, Figure 30. We advise inspecting the slip ring mechanism of your turbine to make sure it is lubricated. We found that Magnalube grease (Magnalube LLC, USA) works very well, especially in the location with saltwater mist (Ocean coast). Interestingly, this grease is not marked as electrically conductive, nevertheless no conductivity issues were encountered. If after some time the turbine does not swivel well, it is time to inspect the slip rings!

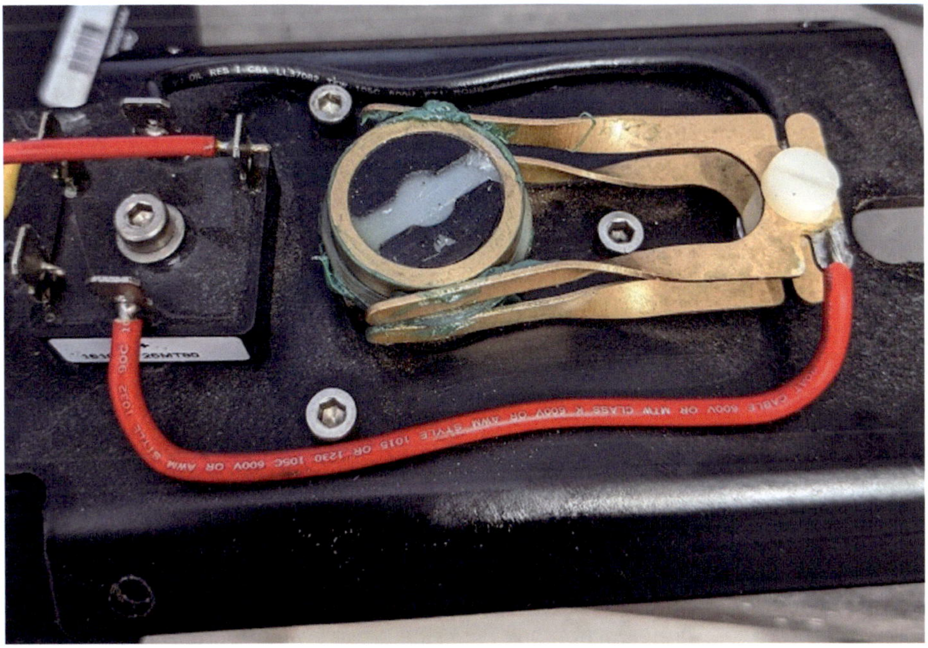

Figure 30: Slip ring mechanism at the base of the turbine. The Magnalube green grease is used for lubricating the contacts. Also notice a 3-phase bridge rectifier on the left.

After ensuring lubrication, it is time to mount the turbine on a mast. One possible option for the mast is a pair of thick walled 5-ft long 1-1/2" galvanized conduit pipes, ID 1.62", OD 1.9" (from McMaster-Carr). The pipes are connected via a tee conduit body, through which the wires exit. In a real-life test, the arrangement being described here withstood wind gusts of 60 mph, which occurred in August 2022 in the San Diego, CA area. Figure 31 shows the mounted turbine along with several accessories, which will be discussed accordingly.

Figure 31: Top, the Chinook turbine mounted onto a mast made of two Rigid conduit pipes joined by a tee; grey junction box houses a brake switch. Bottom left, inside the weatherproof garden box containing the battery, load control circuit, diversion load, and inverter. Bottom right, weatherproof power outlet mounted on the side of the box.

The electrical arrangement is shown in Figure 32. The Chinook 200 turbine provides a built-in full-wave rectification (Figure 30); other turbines may be single or 3-phase AC generators. The pulses must be smoothed by a large capacitance. The resulting DC is managed by a PWM load control circuit. The output of the load control circuit is fed into the battery, which is shunted by a PWM diversion load. The diversion load measures battery voltage and drains excess of charge (via a resistor), when the voltage exceeds a set limit. Charge draining is necessary for protecting the battery from overcharging and keeping the turbine loaded. One cannot simply disconnect the battery, because an unloaded turbine would spin up too fast causing mechanical damage. There are times when the wind or wind gusts are too strong, and it is advisable to disable the turbine. To this end, a switch "Brake" is added to the design that short circuits the turbine. A shorted turbine spins very slowly due to the opposing strong magnetic field.

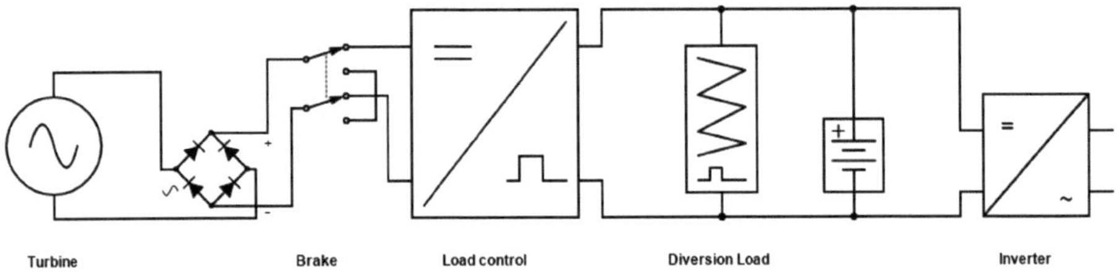

Figure 32: Overall electrical arrangement of the wind turbine setup. From left to right: turbine AC generator, bridge rectifier (built into the turbine), brake switch, load control circuit, diversion load, battery, inverter.

The purpose of load control circuit is to maintain optimal wind energy harvesting by reaching maximum on the Power coefficient – TSR curve, as discussed above, see Figure 28, Figure 29 and ref. 21. The earlier experiments used programmable DC load, which maintained set input voltage and converted energy into heat. The load control circuit is essentially a buck converter [22] that maintains input rather than output voltage. The other very important role of the lad control circuit is to provide flexibility with respect to the voltage of the batteries. The same turbine may charge 12 or 24 V batteries, because it "sees" a specific voltage set by the load control circuit regardless of the batteries.

Figure 33 shows the schematic of the load control circuit as well as an oscillogram of some interesting test points. Full-wave rectified turbine voltage is stored in C1 – a large electrolytic capacitor bank, 3× 270-µF, 200-V rated. The turbine voltage is monitored by the gate driver U1, which has sufficient hysteresis, *i.e.,* turning on at 2.1 to 3.0 V, turning off at 0.8 to 1.5 V. To measure relatively high turbine voltage and to maintain the control window of about 1.5 V, we used a series of Zener diodes, whereby individual diodes may be shorted. Zener diodes effectively subtract a large voltage offset while the reminder appears on R3. Such an arrangement does not require a voltage divider, which would significantly widen the voltage control window. The L2R4C4 network acts as a low pass filter eliminating noise from the turbine voltage for precise measurements. As soon as the turbine voltage reaches the set level, the gate driver U1 turns on, two parallel FETs Q1, Q2 begin conducting and the current through the battery begins rising. The turbine and C1 voltage begin dropping and continue to do so until about 1.5 V below the set level, see the oscillogram at the bottom of Figure 33. After that, the process starts over. The values of C1 and L1 were calculated such that the switching frequency is rather low, hundreds of Hz. The gate driver is powered from a linear regulator 7812, which is in turn powered by a voltage clipper built around Q3. The linear regulator allows max 40 V at the input, which may be too low if we decide to operate the turbine at voltages higher than 40 V. The clipper is a common collector topology that uses a high voltage transistor KSP44, which can sustain 400 V. The clipper limits the voltage around 20 V.

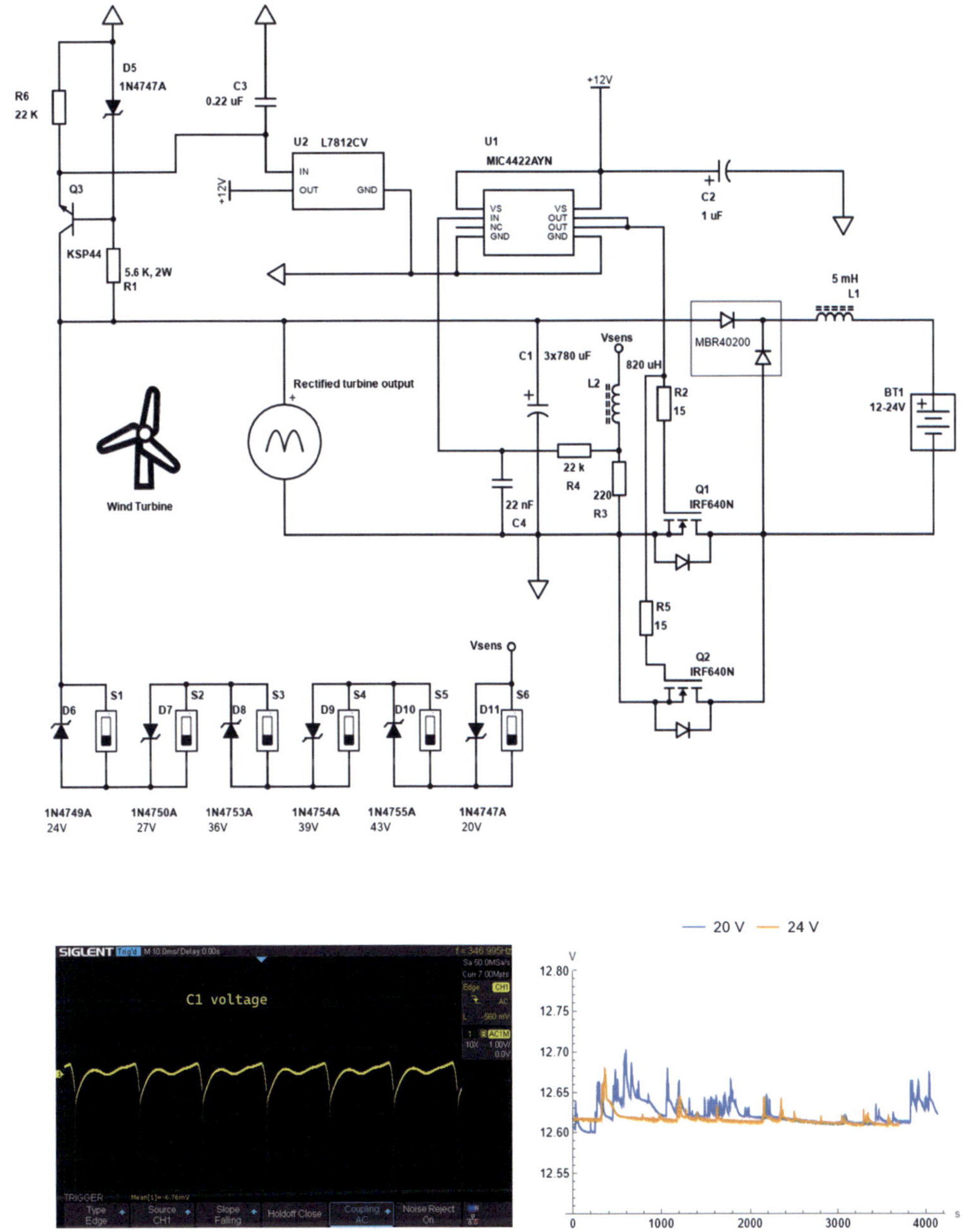

Figure 33: Load control circuit. Top, the "McDonald's" sign indicates full-wave rectified turbine output. DIP switches shorten selected Zener diodes to maintain a desired voltage. Bottom left, an oscillogram of the C1 voltage during the turbine rotation. Bottom right, voltage measured across the battery over ~ 1 hour.

After deploying the turbine into production, optimization steps can be taken. Specifically, voltage setting of the load control circuit can be adjusted to obtain maximum performance of the turbine. Using the Pi-logger, we conducted short-term measurements of the battery voltage using the **voltlog** program (see code in the *Appendix*).

```
nohup sudo ~/voltlog/voltlog -c 1 -a -r 50000 -n 50000 > trial1.txt &
```

The program averages (-a) voltages over 50000 measurements (-n) at the sampling rate 50000 samples/s (-r). Hence, the output occurs every second. To stop the program, a signal must be sent to the process id, for example:

```
sudo kill -s SIGINT 4810
```

Figure 33 bottom right shows measurements at two different load control circuit settings, 20 and 24 V. It seems that at 20 V, more energy transfer occurred. Also, at 24 V, the turbine rotated considerably faster, and a whistling noise was heard. Whistling is a sign of turbulence, which leads to power loss. In all fairness, 1 hour observation per trial is not enough for proper optimization, because the wind speed was likely quite different in the two observation periods. One would have to monitor the performance for a week or so, which is easy to do thanks to the Pi-logger.

Special thanks to my friend and a colleague Michael D. Morris for offering his deck as a platform for the wind generator.

Chapter 4 • Curious Power Sources

There have been proposed several interesting batteries based on microbial transformation of organics, also known as microbial fuel cells. These batteries were based on sediment (2), urine (3, 4), yeast and grape juice (6), and others. These systems rely on the electron transfer from organic matter to oxygen (in the air) mediated by microbes. The electron transfer process is vital for the functioning of any living organism, including the microbes. The circuit consuming the power from the microbial fuel cell *de facto* becomes a part of a microbial electron transfer process (respiration). It was shown that such prototype systems were capable of powering microcontroller-based devices. Imagine turning your compost into a power source in addition to feeding your plants! Successful energy harvesting circuits from the microbial fuel cells have been created (e.g., 5). It is important to distinguish the microbial fuel cells from another "natural" battery such as a potato with iron and copper electrodes. The potato battery is closely related to the original Volta element, whereby two dissimilar metals are immersed into an electrolyte. The potato is that semi-solid electrolyte; but in fact, no potato organic matter participates in the electrochemistry of the energy generation.

For our first experience with renewable energy, let us discuss a microbial fuel cell battery based on rabbit manure. The manure was obtained from our pet rabbit Mr. Flop (Figure 34). Interestingly, it was shown that the rabbit manure could be used to produce biogas due to its rich microbial composition and nutrients (7). The rabbit manure contains cellulose, hemi-cellulose and lignin, which can be digested in our microbial fuel cell by the microbes that are present in the said manure. (Other types of manure will likely work too, because of similar composition, ref. 8.) The manure battery is inspired by the sediment battery (2) with significant modifications. Specifically, unlike the sediment battery, which used sodium acetate as fuel, we will be using manure and hay without any other additional fuels!

Figure 34: Mr. Flop is the "manure engineer" for the microbial fuel cell.

Figure 35 illustrates the steps for making the manure battery. First, prepare 4 gallons (~15 liters) of electrolyte by dissolving 35.4 g of sea salt "instant ocean" in tap water. Assuming that sea salt is mostly NaCl, it makes a 40 mM NaCl solution. The NaCl concentration is chosen based on the work describing the sediment fuel cell (2). Additional electrolytes in the sea salt are likely beneficial to the manure microbes. The other required materials are plastic pail, stainless steel mesh, solid copper wires, terminals, perforated plastic sheet.

Using a roll or a sheet of stainless-steel mesh, cut out appropriately sized cathode and anode electrodes. The size must be such that the electrodes easily fit into the pail covering most of the pail's diameter. Attach a solid copper wire to each electrode using a bolt and a nut. Place one electrode, the anode, on the bottom of the pail. Cover the anode with rabbit manure and hay (biofuel). Cover the biofuel with a perforated plastic disc to prevent hay and other substances from floating up. Pour the electrolyte; place the other electrode, the cathode, such that it is semi-immersed into the electrolyte and exposed to air. Use foam cubes as flotation devices to suspend the cathode. Lastly, short-circuit cathode and anode at the external terminals to enable the "bio-current" to flow while the microbes grow. It is important to understand that although we also have dissimilar metals, such as stainless steel and copper, they do not contribute to the power generation. This is because the dissimilar metals in our case are in a direct contact, hence any voltage generated at the junction is automatically shorted.

Figure 35: Manure battery. Step 1: stainless steel mesh anode placed on the bottom of a plastic pail. Step 2: biofuel (manure and hay) is placed on top of the anode. Step 3: a perforated plastic disc placed on top of the biofuel. Step 4: electrolyte is poured, cathode is situated on the foam floaters.

You will need some patience waiting until the microbial community is actively digesting the biofuel. Consistent with the referenced publications, it will take about a month and a half before any significant output emerges! It is important to keep the **short circuit** between the anode and cathode while the battery is maturing. This is due to the ecosystem evolving and positively selecting the bacteria that can participate in the electrochemical process with current flowing through the external circuit. If the current is not permitted to flow, some other bacteria may grow which are unable to give off the electrons from the biofuel to the stainless-steel anode.

Let's now measure the output of our manure battery. Interesting values are the short-circuit current and the open circuit voltage. The former indicates the amount of material being digested per second. The latter tells us about the nature of chemical species involved in the process. Measuring the short-circuit current is very easy, because its value is stable during the measurement process: the battery is already in the short circuit state, we just replace the jumper wire with a micro amperemeter. Measuring open-circuit voltage is more difficult. It turned out that the voltage keeps increasing over time during the measurement, presumably due to the accumulation of unreduced oxygen. Therefore, we arbitrarily decided to record the voltage after 5 min of open circuit state. Figure 36 shows the current and voltage measurements over a two-month period.

An interesting question to investigate is how much manure and hay to put into the battery? Presumably if we put too much organics, the bacteria may not reach the anode. Hence, there must be an optimal ratio between the amount of the organics and the anode/organics contact area. In addition, the sodium chloride concentration may need to be optimized. Too little salt will harm the bacteria and reduce conductivity of the electrolyte. Too much salt will harm the bacteria by osmotically withdrawing the water.

What can we do with such a low but steady power? One option could be to connect several of such batteries in series to produce usable voltage. Another option is to use an ultra-low voltage energy harvesting circuit based on the LTC3108 integrated circuit (Analog Devices; see datasheet extract reproduced in the Appendix). The device is a pulsed DC-DC converter that operates with voltages as low as 20 mV. An additional important component required is a 1:100 transformer.

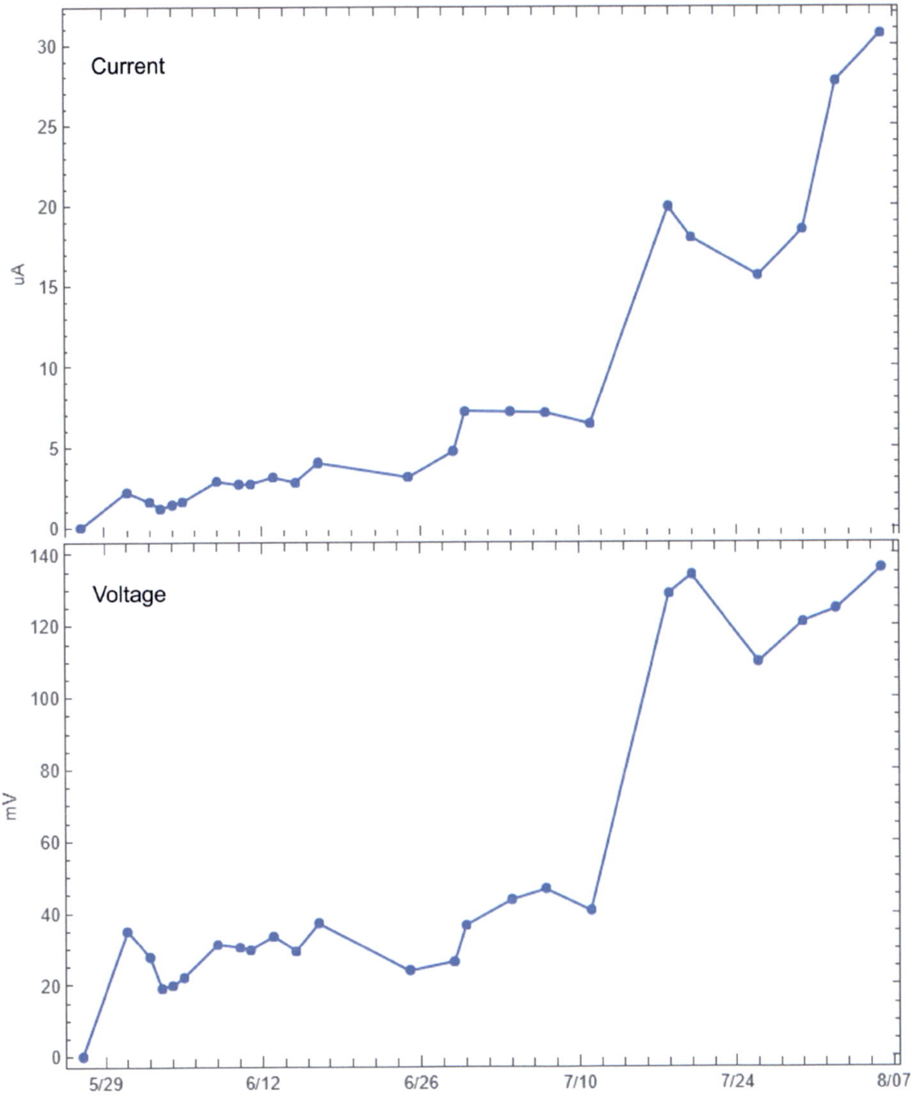

Figure 36: Output of the manure battery. Top, short-circuit current, microamperes; bottom, open circuit voltage after 5 min.

Chapter 5 • Conduits and Batteries

In this chapter we will discuss some boring but critical technical details of our DIY renewable energy power plant. The wiring must be housed in the conduits and the batteries must be taken care of. We do not discuss lithium batteries here because currently they are very "smart". In fact, the large capacity lithium batteries come with a built-in battery management system, which takes care of balancing cells, preventing overcharge, etc. Unlike the lead-acid battery, there is not much that can be done if a lithium battery is degraded. Therefore, for the experimentalist, a lead-acid battery is more interesting!

Conduits

The wiring associated with the PV of the garage power plant, or the deck wind generator is hidden in metallic conduits. The connections are made in the junction boxes. One could choose a non-metallic option with PVC conduits and boxes, but we prefer metal. Among the metallic ones, there are several types of conduits: Electrical Metal Tubing (EMT), Rigid and Flex. Figure 37 shows EMT and Flex conduits as well as the fittings relevant to our project. The EMT is a thin-walled tube, which can be made of galvanized or stainless steel. The Rigid uses much thicker steel; hence it is more expensive but more durable as well. Flex is a metal hose that may be covered with a plastic sheath, in which case it may be used outdoors.

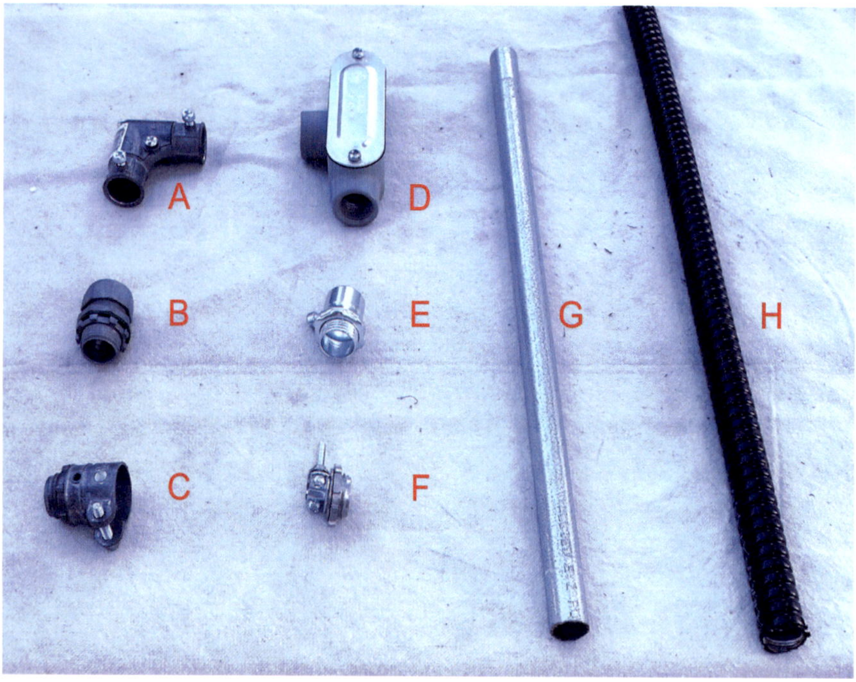

Figure 37: Conduits and fittings. A – two tubes elbow connector, indoors use. B – raintight tube to box compression connector, in/outdoor use. C – Flex to box connector, indoors use. D – tee conduit body to join 3 tubes. E – set screw tube to box connector, indoors use. F – cable to box screw clamp, indoor use. G – EMT conduit. H – plastic-covered Flex conduit.

A good starting point for the wiring is the building inlet for the PV or wind power. Place a junction box at this place. Establish the conduit runs in the building and use conduit bodies for turns. The wires of appropriate gauge may be solid or stranded.

A steel "fish tape" is very handy for pulling a bundle of wires through a conduit. Insert a fish tape into one end of the conduit at the junction box or a conduit body, push it forward until it shows up on the other end, such as another box. Tie a "pulling line", *i.e.,* a nylon rope to the hook of the fish tape and pull everything back. Detach the pulling line from the fish tape. At this point, the pulling line is going through the conduit, and it is dangling on both ends. Tie the end of the bundle of wires to the pulling line using electric tape. Carefully pull on the rope to draw the wires through the conduit.

Lead-acid Batteries

The battery capacity can be assessed with Pi-logger. For example, Figure 38 shows charging and discharging dynamics of the batteries when we suspected diminished battery capacity. The current is characteristic of the entire bank. The voltage was measured only on one battery, however given the battery equalizer, all other batteries had a similar voltage. Apparently, in the "dark time", *i.e.,* from about 17:00 until 10:00 next day (~17 h), the batteries were discharged at 1 A reaching a complete discharge. This gives us about 17 Ah, instead of the expected 80 Ah.

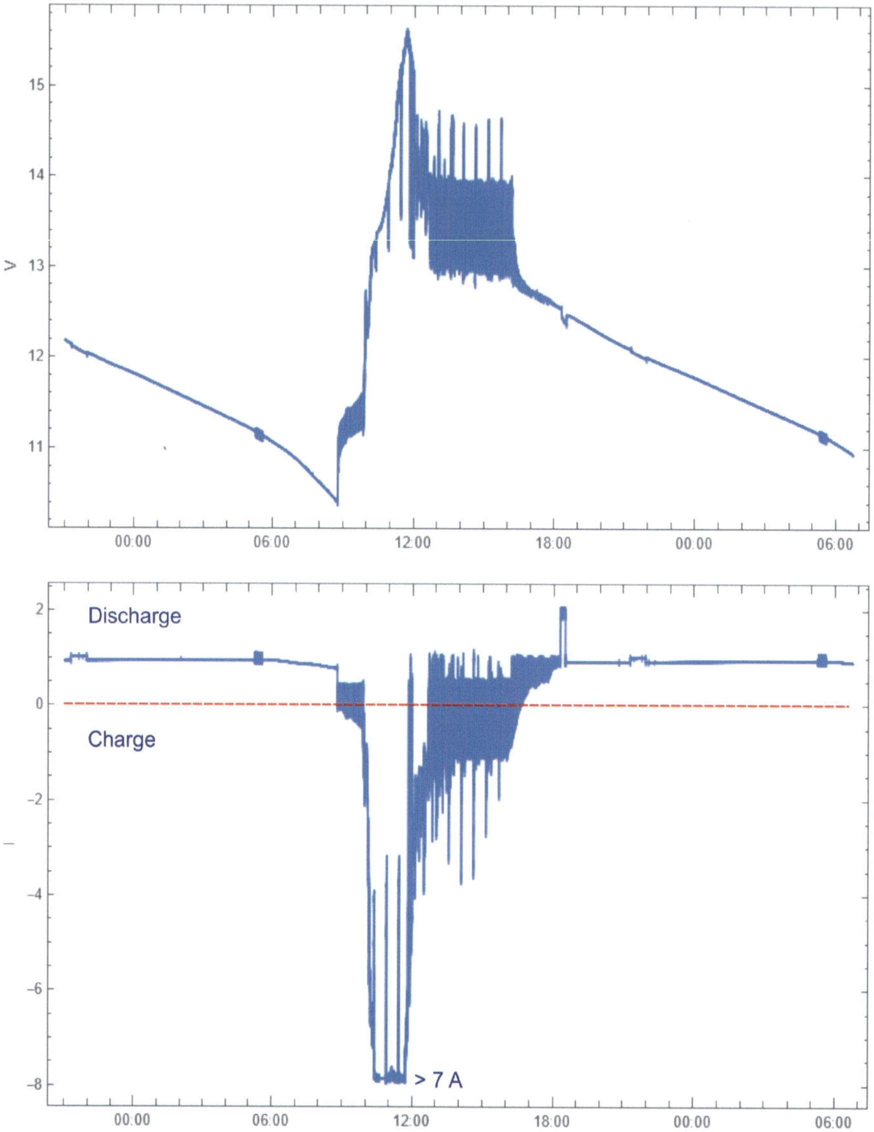

Figure 38: An example of a charging and discharging cycle of the battery bank. The voltage measured on one of the batteries, the current is characteristic of the entire bank.

Using the "old school" non-sealed lead-acid batteries has the advantage of diagnosing and potentially recovering the batteries. Figure 39 shows our "battery clinic" where we measure the electrolyte (sulfuric acid) density and replace the electrolyte if necessary.

The charging process of a 12 V lead-acid battery involves three plus one stages. The *bulk* stage keeps the charging current (high) constant until battery reaches 14.4 to 14.8 V. The current will reduce while maintaining the fixed voltage leading to the *absorption* stage. At the absorption, the voltage is reduced to 14.2 to 14.4 V to top off the charge (about 30%). The current further reduces as the battery becomes fully charged. Finally, the third *float* stage begins. The voltage at this stage is reduced to 13.2 to 13.8 V passing a relatively small current compensating for the battery self-discharge. After several hours of floating, the electrolyte density will reach a certain value indicative of a full charge. Approximately once a month it is useful to run the equalization charge, at which the voltage is held around 15.5 to 16 volts. Equalization helps to prevent the buildup of stratification, where acid concentration is higher at the bottom of the battery. During that time one can hear the batteries "boil" producing hydrogen. No smoking or barbequing around the lead acid batteries!

Despite charging and waiting, the density of the electrolyte in our batteries would not increase. We decided to replace the electrolyte. The electrolyte and the hydrometer can be purchased in an automotive store rather cheaply. Usually, the hydrometer indicates the state of charge along with the value of the electrolyte density. Every cell of the battery must be evaluated in terms of the electrolyte density. Generally, electrolyte is likely to be replaced in all cells.

Wearing the proper attire of a "battery doctor", *i.e.*, gloves, safety glasses, and a coat, open the cells with a screwdriver prying off the plugs. Aspirate some quantity of the fluid, assess the density, and return the fluid back to the cell. Keeping the cells open, carefully tilt the battery on a solid support (e.g., cinder block) and tilt to pour out the electrolyte. Rock the battery back and forth to remove the electrolyte completely. Bring the battery close to the electrolyte container and fill up the cells using a hose attached to the electrolyte container. Fill the fluid until it shows up just below the opening of the cells. Cap the cells and bring the battery back to service. Dispose of the drained electrolyte according to the local hazmat code.

Figure 39: Diagnostics and revival of lead-acid batteries. Top left, the battery clinic set up: container to receive old electrolyte; cinder block; new electrolyte; a screwdriver; hydrometer. Top right: measuring the electrolyte density. Bottom left: pouring out old electrolyte. Bottom right: filling up cells with the new electrolyte.

Literature

1. Glover TJ. Pocket Ref. SequoiaPocket_ref Publishing Inc, USA.

2. Thomas YRJ, Picot M, Carer A, Berder O, Sentieys O, Barriere F. A singleThomas_13 sediment-microbial fuel cell powering a wireless telecommunication system. JOURNAL OF POWER SOURCES, 2013, Vol 241, 703-708.

3. Walter XA, Gajda I, Forbes S., et al. Scaling-up of a novel, simplified MFC stack basedWalter_16 on a self-stratifying urine column. BIOTECHNOLOGY FOR BIOFUELS, 2016, Vol 9, art. no 93.

4. Walter XA, Greenman J, Ieropoulos IA. Microbial fuel cells directlyWalter_20 powering a microcomputer. JOURNAL OF POWER SOURCES, 2020, Vol 446, art. no 227328.

5. Yamashita T, Hayashi T, Iwasaki H, et al. Ultra-low-power energy harvester for microbial fuel cells and its application Yamashita_19to environmental sensing and long-range wireless data transmission. JOURNAL OF POWER SOURCES, 2019, Vol 430, 1-11.

6. de Oliveira AHP, Alcaraz-Espinoza JJ, da Costa MM, Nascimento MLF, Swager TM, de Oliveira HP. ImprovementOlivera_19 of Baker's yeast-based fuel cell power output by electrodes and proton exchange membrane modification. Materials Science and Engineering: C, 2019, 105, 110082.

7. Adrover ME, Cotabarren I, Madies E, Rayes M, Reartes SBR, Pedernera M. Anaerobic co-digestion of rabbitAdrover_20 manure and sorghum crops in a bench-scale biodigester. Bioresour. Bioprocess, 2020, 7:42.

8. Li K, Liu R, Sun C. Comparison of anaerobic digestion characteristics and kinetics of four livestock manures with differentLi_15 substrate concentrations. Bioresource Technology, 2015, 198, 133–140.

9. MIT freeMITocw courses. https://ocw.mit.edu/

10. PV educationpved https://www.pveducation.org/.

11. Becquerel E. «Mémoire sur les effetsBecquerel_1839 électriques produits sous l'influence des rayons solaires». Comptes Rendus, 1839, Vol 9, 561–567.

12. Henry CH. Limiting efficiencies of idealHenry_80 single and multiple energy gap terrestrial solar cells. Journal of Applied Physics, 1980, 51, 4494.

13. Polman A, Knight M, Garnett EC, Ehrler B, Sinke WC. PhotovoltaicPolman_16 materials: Present efficiencies and future challenges. SCIENCE, 2016, Vol 352, ISSUE 6283.

14. Gueymard CA, Myers D, Emery K. Proposed referenceGueymard_02 irradiance spectra for solar energy systems testing. Solar Energy, 2002, Vol 73, 443–467.

15. Hallwachs W. Über den EinflussHallwachs_1888 des Lichtes auf electrostatisch geladene Körper. Annalen der Physik. 1888, 269: 301–312.

16. Malathy S, Ramaprabha R. Reconfiguration strategies to extract maximum power from photovoltaic array under partially shadedMalathy_18 conditions. Renewable and Sustainable Energy Reviews, 2018, 81, 2922-2934.

17. Riedijk FR, Huijsing JH. A smart balanced thermal pyranometer using a sigma-delta A-to-D converter for direct communicationRiedijk_93 with microcontrollers. Sensors and Actuators A: Physical, 1993, 37–38, 16-25.

18. Engineering ToolBox, (2003). Emissivity CoefficientsEngToolbx common Products.

19. J. F. Manwell, J. G. McGowan and A. L. Rogers. Wind Energy Explained: Manwell_09Theory, Design and Application, Second Edition 2009. John Wiley & Sons.

20. Elizondo J, Martinez J and Probst O. ExperimentalElizondo_09 study of a small wind turbine for low- and medium-wind regimes. Int. J. Energy Res. 2009, 33:309–326

21. Pozhitkov A. Taming YourPozitkov_19 Wind Turbine. Circuit Cellar. 2019, #347

22. Brown M, Kularatna N, Mack RA, Maniktala S. (2008). PowerBuckConverter sources and supplies: World class designs. Elsevier Science & Technology.

Appendix

Program: Voltread

```
#include "daqhats_utils.h"
#include <daqhats/daqhats.h>
#include <unistd.h>


void prompt()
{
  fprintf(stderr, "\nX (q to stop): ");
}

double * MatrixElement(uint32_t r, uint32_t c, double * data, uint32_t dims[2])
{
  if(r < dims[0] && c < dims[1])
    return &data[r * dims[1] + c];
  return NULL;
}

int main(int argc, char *argv[])
{
    uint8_t address;
    int channelmask = 1;
    int numread = 10;
    uint32_t options = OPTS_DEFAULT;
    double scanrate = 0.0;

    int opt;
    while ((opt = getopt(argc, argv, "c:r:n:")) != -1) {
    switch (opt) {
     case ‘c’:
         channelmask = atoi(optarg);
         break;
     case ‘n’:
         numread = atoi(optarg);
         break;
     case ‘r’:
         scanrate = atof(optarg);
         break;

     default: /* ‘?’ */
         fprintf(stderr, "Usage: %s [-c channel_mask] [-r sampling_rate] [-n
samples_to_read]\n Defaults: channel_mask=1, samplig_rate=MAX, samples_to_
read=10;", argv[0]);
```

```
            return -1;
            }
    }

    char display_string[256] = "";
    int result = RESULT_SUCCESS;
    int mcc118_num_channels = mcc118_info()->NUM_AI_CHANNELS;
    int num_channels = 0;

    { // determine the number of channels
      int cm = channelmask;
      while (cm > 0)
      {
          if (cm & 0x01)
            num_channels++;
          cm >>= 1;
      }
    }

    if(num_channels > mcc118_num_channels)
    {
      fprintf(stderr, "\nRequested number of channels %d exceeds available %d",
num_channels, mcc118_num_channels);
      return -1;
    }

// Determine the address of the device to be used
    if (select_hat_device(HAT_ID_MCC_118, &address) != 0)
    {
        return -1;
    }

// Open a connection to each device
    result = mcc118_open(address);
    STOP_ON_ERROR(result);

    convert_options_to_string(options, display_string);
    fprintf(stderr, "    Options: %s\n", display_string);

    if(scanrate == 0.0)
      scanrate = 100000.0/num_channels;
    double actual_scan_rate = 0.0;
    mcc118_a_in_scan_actual_rate(num_channels, scanrate, &actual_scan_rate);
    fprintf(stderr, "\nRequested %f and actual %f scan rates.\nPress enter to
stop...\n", scanrate, actual_scan_rate);
```

```
    uint32_t buffer_size = numread * num_channels;
    double *read_buf = (double *) malloc(buffer_size * sizeof(double));

    uint32_t matrixdims[2];
    prompt();
    double x;
    while(scanf(«%lf», &x)!=0)
    {
      uint16_t status;
      uint32_t samples_per_channel=0;

      result = mcc118_a_in_scan_start(address, channelmask, numread, scanrate,
options);  // fixed nuber of samples
      STOP_ON_ERROR(result);

      matrixdims[1] = num_channels;
      do{
        result = mcc118_a_in_scan_read(address, &status, numread, -1, read_buf,
buffer_size, &samples_per_channel);
        STOP_ON_ERROR(result);

        if (status & STATUS_HW_OVERRUN)
        {
            fprintf(stderr, "\n\nHardware overrun\n");
            break;
        }
        else if (status & STATUS_BUFFER_OVERRUN)
        {
            fprintf(stderr, "\n\nBuffer overrun\n");
            break;
        }

        matrixdims[0] = samples_per_channel;
        //average per column
        printf("\n%lf\t",x);
        for(int col = 0; col < num_channels && samples_per_channel > 0; col++)
        {
          double colaverage = 0;
          for(int row = 0; row < samples_per_channel; row++)
             colaverage += *MatrixElement(row, col, read_buf, matrixdims);
          colaverage /= samples_per_channel;
          printf("%lf\t", colaverage);
        }
      } while(status & STATUS_RUNNING);

      result = mcc118_a_in_scan_cleanup(address);
```

```
        STOP_ON_ERROR(result);
        prompt();
    }
stop:
    result = mcc118_close(address);
    print_error(result);
    free(read_buf);
    return 0;
}
```

Program: Voltlog

```
#include "daqhats_utils.h"
#include <daqhats/daqhats.h>
#include <unistd.h>
#include <signal.h>

#define MAX_SCANRATE 90000.0

int gContinue=1;

double * MatrixElement(uint32_t r, uint32_t c, double * data, uint32_t dims[2])
{
  if(r < dims[0] && c < dims[1])
    return &data[r * dims[1] + c];
  return NULL;
}

void sigIntHandler(int signum) {
    gContinue = 0;
}

int main(int argc, char *argv[])
{
    signal(SIGINT, sigIntHandler);
    uint8_t address;
    int channelmask = 1;
    int numread = 10;
    uint32_t options = OPTS_DEFAULT;
    double scanrate = 0.0;
    int averaging = 0;

    int opt;
    while ((opt = getopt(argc, argv, "c:r:n:a")) != -1) {
    switch (opt) {
```

```
        case 'c':
            channelmask = atoi(optarg);
            break;
        case 'n':
            numread = atoi(optarg);
            break;
        case 'r':
            scanrate = atof(optarg);
            break;
        case 'a':
            averaging = 1;
            break;

        default: /* '?' */
            fprintf(stderr, "Usage: %s [-c channel_mask] [-r sampling_rate] [-n
samples_per_batch] [-a]\n Defaults: channel_mask=1, samplig_rate=90000/number_of_
channels per sec., samples_per_batch=10, averaging per batch off", argv[0]);
            return -1;
            }
    }

    char display_string[256] = "";
    int result = RESULT_SUCCESS;
    int mcc118_num_channels = mcc118_info()->NUM_AI_CHANNELS;
    int num_channels = 0;

    { // determine the number of channels
      int cm = channelmask;
      while (cm > 0)
      {
          if (cm & 0x01)
            num_channels++;
          cm >>= 1;
      }
    }

    if(num_channels > mcc118_num_channels)
    {
      fprintf(stderr, "\nRequested number of channels %d exceeds available %d",
num_channels, mcc118_num_channels);
      return -1;
    }

// Determine the address of the device to be used
    if (select_hat_device(HAT_ID_MCC_118, &address) != 0)
    {
```

```
        return -1;
    }

// Open a connection to each device
    result = mcc118_open(address);
    STOP_ON_ERROR(result);

    convert_options_to_string(options, display_string);
    fprintf(stderr, "    Options: %s\n", display_string);

    if(scanrate == 0.0)
      scanrate = MAX_SCANRATE/num_channels;
    double actual_scan_rate = 0.0;
    mcc118_a_in_scan_actual_rate(num_channels, scanrate, &actual_scan_rate);
    fprintf(stderr, "\nRequested %f and actual %f scan rates.\n", scanrate,
actual_scan_rate);

    uint32_t buffer_size = numread * num_channels;
    double *read_buf = (double *) malloc(buffer_size * sizeof(double));

    uint32_t matrixdims[2];
    matrixdims[1] = num_channels;
    uint16_t status;
    uint32_t samples_per_channel=0;
    while(gContinue)
    {
      result = mcc118_a_in_scan_start(address, channelmask, numread, scanrate,
options);  // fixed nuber of samples
      STOP_ON_ERROR(result);

        matrixdims[0] = samples_per_channel;
        if(averaging)
        {
        //average per column
          printf("\n");
          for(int col = 0; col < num_channels && samples_per_channel > 0; col++)
          {
            double colaverage = 0;
            for(int row = 0; row < samples_per_channel; row++)
              colaverage += *MatrixElement(row, col, read_buf, matrixdims);
            colaverage /= samples_per_channel;
            printf("%lf\t", colaverage);
          }
        } else {
          for(int row = 0; row < samples_per_channel; row++)
          {
```

```
            printf("\n");
            for(int col = 0; col < num_channels && samples_per_channel > 0;
col++)
              printf("%lf\t",*MatrixElement(row, col, read_buf, matrixdims));
          }
        }

      do{
        result = mcc118_a_in_scan_read(address, &status, numread, -1, read_buf,
buffer_size, &samples_per_channel);
        STOP_ON_ERROR(result);

        if (status & STATUS_HW_OVERRUN)
        {
            fprintf(stderr, "\n\nHardware overrun\n");
            break;
        }
        else if (status & STATUS_BUFFER_OVERRUN)
        {
            fprintf(stderr, "\n\nBuffer overrun\n");
            break;
        }

      } while(status & STATUS_RUNNING);

      result = mcc118_a_in_scan_cleanup(address);
      STOP_ON_ERROR(result);

   }
stop:
    result = mcc118_close(address);
    print_error(result);
    free(read_buf);
    fprintf(stderr, "\nLogging ended\n");
    return 0;
}
```

Hybrid inverter

MN5048DIY
User Manual

1.2 Basic system introduction

The figure below shows the system application scenario of this product.
A complete system consists of the following parts:

1. PV module: Converts light energy into DC power, and charges the battery through the onboard MPPT solar charge controller, or to directly drive the inverter making AC power to drive the loads. (battery-less mode)

2. Mains or generator: Connected at the AC input, to power the loads while charging the battery. If the mains or generator is not connected, the system can also operate normally, and the load is powered by the battery and PV array.

3. Battery: Provided to ensure normal power supply to the system loads when solar energy is insufficient and the AC Main is not connected.

4. Household load: Allow connection of various household and office loads, including refrigerators, lamps, TVs, fans and air conditioners.

5. MNS Hybrid Inverter: The energy conversion unit of the whole system. Specific system wiring method depends on the actual application scenario.

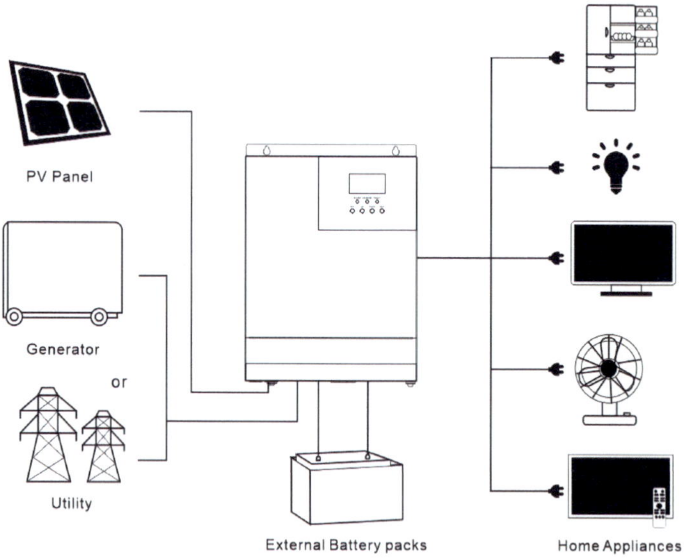

10-539-1 MN5048DIY REV - 5 of 39

MN5048DIY
User Manual

Parameter no.	Parameter name	Settings	Description
00	Exit setting menu	[00] ESC	Exit the setup menu
01	Output source priority	[01] SOL	PV priority mode, switching to the Mains when the PV fails or the battery is lower than the set value of parameter [04].
		[01] UTI default	Mains priority mode, switching to inverter only when the mains fail .
		[01] SBU	Inverter priority mode, switching to the mains only when the battery is under voltage or lower than the set value of parameter [04].
02	Output Frequency	[02] 50.0	Bypass self-adaptation; when the mains are connected, it automatically adapts to the mains frequency; when the mains iare disconnected, the output frequency can be set through this menu. The default output frequency of the 230V machine is 50HZ, and the 120V machine 60HZ.
		[02] 60.0	
03	AC Input Voltage Range	[03] APL	Wide mains input voltage range of 230V machine: 90~280V Mains input voltage range of 120V machine: 90~140V
		[03] UPS default	Narrow mains input voltage range of 230V machine: 170~280V Mains input voltage range of 120V machine: 90~140V
04	Battery Power to Utility Setpoint	[04] 46.0 default	When the parameter [01] =SOL/SBU, the battery voltage is lower than the set value, and the output is switched from the inverter to the mains. Setting range: 44V - 52V.
05	Utility to Battery Power Setpoint	[05] 56.0V default	When the parameter [01] =SOL/SBU, the battery voltage is higher than the set value, and the output is switched from the mains to the inverter. Setting range: 48V~60V.

MN5048DIY
User Manual

Parameter no.	Parameter name	Settings	Description
06	Charger source priority	[06] CSO	PV priority charging; when not PV is unavailable, charging from the mains is started.
		[06] CUB	Mains priority charging; When charging from the mains is not available PV charging is started.
		[06] SNU default	PV and Mains hybrid charging; PV charging is a priority, and when the PV energy is insufficient, the Mains charging supplements. When the PV energy is sufficient, the Mains charging stops. Note: Only when the Mains bypass output is loaded, the PV charging and the mains charging can work at the same time. When the inverter is in use, only the PV charging can be started.
		[06] OSO	Only PV charging, with the Mains charging not activated.
07	Max charger current	[07] 60A default	Max charger current (AC charger+PV charger). Setting range: 0~80A
08	Battery Type	[08] USE	User-defined; all battery parameters can be set.
		[08] SLd	Sealed lead-acid battery; absorption charge Voltage: 57.6V, float charge voltage: 55.2V
		[08] FLd	Vented lead-acid battery; absorption charge voltage: 58.4V, float charge voltage: 55.2V
		[08] GEL default	Colloidal lead- acid battery; absorption charge voltage: 56.8V float charge voltage: 55.2V
		[08] LL14/L15/L16	Lithium iron phosphate battery L14/L15/L16 corresponds to lithium iron phosphate battery 14 strings/15 strings/16 strings;16 string/15 string/14 string default constant The charging voltage is 56.8V, 53.2V, 49.6V, which are adjustable.

MN5048DIY
User Manual

Parameter no.	Parameter name	Settings	Description
		[08] N13/N14	Ternary lithium battery; which is adjustable.
09	Battery Absorption charge voltage	[09] 57.6V default	Boost (absorb) charge voltage setting; the setting range is 48~58.4V with a step of 0.4V for user-defined battery and lithium battery.
10	Battery Absorption charge time	[10] 120 default	Boost (absorb) charge maximum time setting, which means the maximum charging time to reach the set voltage of parameter [09] during constant-voltage charging. The setting range is 5min~900min, with a step of 5 minutes. It is valid for user-defined battery and lithium battery.
11	Battery float charge voltage	[11] 55.2V default	Float charge voltage, setting range: 48V~58.4V at 0.4V step is valid in for a user-defined battery.
12	Battery over discharge voltage (delay off)	[12] 42V default	Over-discharge voltage ; when the battery voltage drops to the set point, after the delay time set by parameter [13] the inverter will stop inverting. Setting range is 48V - 58.4V with a step of 0.4V. It is valid for user - defined battery and lithium battery.
13	Battery over discharge delay time	[13] 5S default	Over-discharge delay time; when the battery voltage is lower than the parameter [12], the inverter output will be turned off after the time set by this parameter is delayed. The setting range is 5S~55S, with a step of 5S. It is valid for user-defined battery and lithium battery.
14	Battery under voltage alarm	[14] 44V default	Battery undervoltage alarm point ; when the battery voltage is lower than the point, an undervoltage alarm is given , and the output is not turned off; the setting range is 48V ~ 52V with a step of 0.4V. It is valid for user-defined battery and lithium battery .

MidNite Solar, Inc

MN5048DIY User Manual

Parameter no.	Parameter name	Settings	Description
15	Battery discharge limit voltage	[15] 40V default	Battery discharge limit voltage; when the battery voltage is lower than the point, the output is turned off immediately; the setting range is 40V~ 52V, with a step of 0.4V. It is valid for user -defined battery and lithium battery.
16	Battery equalization enable	[16] DIS	Equalizing charge is disabled
		[16] ENA default	Equalizing charge is enabled, only valid for vented lead-acid battery and sealed lead-acid battery
17	Battery equalization voltage	[17] 58.4V default	Equalizing charge voltage; setting range: 48V~ 58.4V, with a step of 0.4V; valid for vented lead-acid battery and sealed lead-acid battery
18	Battery equalized time	[18] 120 default	Equalizing charge time ; setting range : 5min~900min, with a step of 5 minutes; valid for vented lead-acid battery and sealed lead-acid battery
19	Battery equalized time out	[19] 120 default	Equalizing charge delay; setting range : 5min~900min, with a step of 5 minutes ; valid for vented lead-acid battery and sealed lead-acid battery
20	Battery equalization interval	[20] 30 default	Equalizing charge derating time, 0~30days, with a step of 1 day; valid for vented lead-acid battery and sealed lead-acid battery
21	Battery equalization immediately	[21] DIS	Stop equalizing charge immediately.
		[21] ENA default	Start equalizing charge immediately.
22	Power saving mode	[22] DIS default	Power saving mode disabled .
		[22] ENA	When the power saving mode is enabled, if there is no load or less than 50W, the inverter output is turned of f after a delay for a certain period of time. When the loadis more than 50W, the inverter automatically restarts.
23	Restart when over load	[23] DIS	Automatic restart when overload is disabled. If an overload occurs and the output is turned off, the unit will not restart .

MN5048DIY User Manual

Parameter no.	Parameter name	Settings	Description
24	Restart when over temperature	[24] DIS	Auto matic restart when over temperature is disabled. If an over-temperature shutdown occurs, machine will not restart to turn the output on.
		[24] ENA default	Automatic restart when over temperature is enabled. If an over-temperature shutdown occurs, the machine will restart when the temperature drops.
25	Alarm enable	[25] DIS	Alarm is disabled
		[25] ENA default	Alarm is enabled
26	Beeps while primary source is interrupted	[26] DIS	Alarm beep is disabled when the status of the main input source changes
		[26] ENA default	Alarm beep is enabled when the status of the main input source changes
27	Bypass output when over load	[27] DIS	It is disabled to automatically switch to the Mains when the inverter is overloaded.
		[27] ENA default	It is enabled to aut omatically switch to the Mains when the inverter is overloaded.
28	Max AC charger current	[28] 40A default	Max AC charger current. Setting range: 0~40A ;
29	Split Phase	[29] DIS default	Supply for industrial frequency transformer (disabled)
		[29] ENA	Supply for industrial frequency transformer (enabled)
30	Rs485 address setting	[35] 1 default	Rs485 communication address setting range 1 ~ 254, (refer to Number [32] is valid when set as SLA)
35	Battery undervoltage recovery point	[35] 50.4V default	When the battery voltage is under voltage, the battery voltage needs to recover more than this set value before the inverter starts the output
36	Max PV charger current	[36] 40A default	Max PV charger current. Setting range: 0~40A
37	Battery fully charged recovery point	[37] 52V default	After the battery is fully charged, it needs to be lower than this set voltage before it can be recharged

MN5048DIY
User Manual

4.3 Battery type parameters

For Lead - acid Battery :

Parameters \ Battery type	Sealed lead acid battery (SLD)	Colloidal lead acid battery (GEL)	Vented lead acid battery (FLD)	User - defined (User)
Overvoltage disconnection voltage	60V	60V	60V	36 ~ 60V
Equalizing charge voltage	58.4V	56.8V	59.2V	36 ~ 60V
Absorption charge voltage	57.6V	56.8V	58.4V	36 ~ 60V (Adjustable)
Float charge voltage	55.2V	55.2V	55.2V	36 ~ 60V (Adjustable)
Undervoltage alarm voltage	44V	44V	44V	36 ~ 60V (Adjustable)
Low voltage disconnection voltage	42V	42V	42V	36 ~ 60V (Adjustable)
Discharge limit voltage	40V	40V	40V	36 ~ 60V (Adjustable)
Over - discharge delay time	5s	5s	5s	1 ~ 30s (Adjustable)
Equalizing charge duration	120 minutes	-	120 minutes	0 ~ 600 minutes (Adjustable)
Equalizing charge interval	30 days	-	30 days	0 ~ 250 days (Adjustable)
Absorption charge duration	120 minutes	120 minutes	120 minutes	10 ~ 600 minutes (Adjustable)

MN5048DIY
User Manual

For Lithium Battery :

Battery type / Parameters	Ternary lithium battery (N 13)	Ternary lithium battery (N 14)	Lithium iron phosphate battery (LF16)	Lithium iron phosphate battery (LF15)	Lithium iron phosphate battery (LF14)	User-defined (User)
Overvoltage disconnection voltage	60V	60V	60V	60V	60V	36 ~ 60V
Absorption charge voltage	53.2V (Adjustable)	57.6V (Adjustable)	56.8V (Adjustable)	53.2V (Adjustable)	49.2V (Adjustable)	36 ~ 60V (Adjustable)
Floating charge voltage	53.2V (Adjustable)	57.6V (Adjustable)	56.8V (Adju stable)	53.2V (Adjustable)	49.2 (Adjustable)	36 ~ 60V (Adjustable)
Undervoltage alarm voltage	43.6V (Adjustable)	46.8V (Adjustable)	49.6V (Adjustable)	46.4V (Adjustable)	43.2V (Adjustable)	36 ~ 60V (Adjustable)
Low voltage disconnection voltage	38.8V (Adjus table)	42V (Adjustable)	48.8V (Adjustable)	45.6V (Adjustable)	42V (Adjustable)	36 ~ 60V (Adjustable)
Discharge limit voltage	36.4V	39.2V	46.4V	43.6V	40.8V	36 ~ 60V (Adjustable)
Over-discharge delay time	30s (Adjustable)	30s (Adjustable)	30s (Adjustable)	30s (Adjustable)	30s (Adjustable)	1 ~ 30s (Adjustable)
Absorption charge duration	120 minutes (Adjustable)	120 minutes (Adjustable)	120 minu tes (Adjustable)	120 minutes (Adjustable)	120 minutes (Adjustable)	10 ~ 600 minutes (Adjustable)

8.Technical Parameter

Model	MN5048DIY
AC mode	
Rated input voltage	110/120Vac
Input voltage range	(90Vac-140Vac)
Frequency	50Hz/ 60Hz (auto-sensing)
Frequency range	47±0.3Hz~55±0.3Hz(50Hz); 57±0.3Hz~65±0.3Hz(60Hz);
Overload/short-circuit protection	Breaker
Efficiency	>95%
Conversion time (Bypass and inverting)	10ms(Typicalvalue)
AC reverse flowing protection	yes
Maximum bypass overload current	63A
Inverting mode	
Output voltage waveform	Pure sine wave
Rated output power(VA)	5000(4100/ 4300/4500)
Rated output power(W)	5000(4100/ 4300/4500)
Power factor	1
Rated output voltage(Vac)	120Vac(100/105/110Vac Settable)
Output voltage error	±5%
Output frequency range(Hz)	50Hz±0.3Hz 60Hz±0.3Hz
Efficiency	>90%
Overload protection	(102%<load<110%) ±10%: reporting error and turn off the output after 5 minutes; (110%<load<125%) ±10%: reporting error and turn off the output after 10 seconds; Load>125% ±10%: reporting error and turn off the output after 5 seconds;

MN5048DIY
User Manual

Peak power	10000VA
Loaded motor capacity	4HP
Output short-circuit protection	Breaker
Specification of bypass breaker	63A
Rated battery input voltage	48V (minimum start voltage 44V)
Battery voltage range	40.0Vdc~60Vdc ±0.6Vdc (undervoltage alarm/turnoff voltage/overvoltage alarm/overvoltage restoration...settable LCDscreen)
Power saving mode	Load ≤ 50W
AC charge	
Battery type	Lead acid or lithium battery
Maximum charge current	40A
Charge current error	±5Adc
Charge voltage range	40–60Vdc
Breaker specification	63A
Overcharge protection	Turn off charge after 1min alarm
Solar charge	
Maximum PV open -circuit voltage	500Vdc
PV operation voltage range	120-500Vdc
MPPT voltage range	120-450Vdc
Battery voltage range	40-60Vdc
Maximum output power	5000W
Charge current range of solar energy(settable)	0-80A
Charge short-circuit protection	Blown fuse
Wiring protection	Inverse wiring protection

MN5048DIY
User Manual

Authentication specification	
Specification authentication	CE(IEC62109)、RoHs
EMC authentication grade	En61000
Operation temperaturerange	-15°Cto55°C
Storage temperaturerange	-25°C~60°C
Humidity range	5% to 95% (three-proof paint protection)
Noise	≤60dB
Thermal dissipation	Forced cooling with adjustable air speed
Communication interface	USB/RS485(WiFi/GPRS)/dry node control
Dimension(L*W*D)	426mm*322mm*126mm
Weight(kg)	11.6

DS18B20

PRELIMINARY

DS18B20
Programmable Resolution 1-Wire® Digital Thermometer

DALLAS SEMICONDUCTOR

www.dalsemi.com

FEATURES

- Unique 1-Wire interface requires only one port pin for communication
- Multidrop capability simplifies distributed temperature sensing applications
- Requires no external components
- Can be powered from data line. Power supply range is 3.0V to 5.5V
- Zero standby power required
- Measures temperatures from -55°C to +125°C. Fahrenheit equivalent is -67°F to +257°F
- ±0.5°C accuracy from -10°C to +85°C
- Thermometer resolution is programmable from 9 to 12 bits
- Converts 12-bit temperature to digital word in 750 ms (max.)
- User-definable, nonvolatile temperature alarm settings
- Alarm search command identifies and addresses devices whose temperature is outside of programmed limits (temperature alarm condition)
- Applications include thermostatic controls, industrial systems, consumer products, thermometers, or any thermally sensitive system

PIN ASSIGNMENT

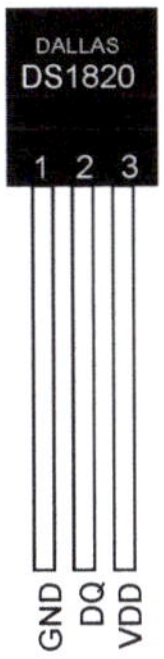

BOTTOM VIEW

DS18B20 To-92 Package

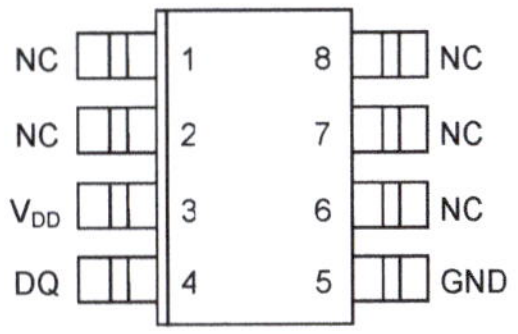

DS18B20Z
8-Pin SOIC (150 mil)

PIN DESCRIPTION

GND - Ground
DQ - Data In/Out
V_{DD} - Power Supply Voltage
NC - No Connect

DESCRIPTION

The DS18B20 Digital Thermometer provides 9 to 12-bit (configurable) temperature readings which indicate the temperature of the device.

Information is sent to/from the DS18B20 over a 1-Wire interface, so that only one wire (and ground) needs to be connected from a central microprocessor to a DS18B20. Power for reading, writing, and performing temperature conversions can be derived from the data line itself with no need for an external power source.

Because each DS18B20 contains a unique silicon serial number, multiple DS18B20s can exist on the same 1-Wire bus. This allows for placing temperature sensors in many different places. Applications where this feature is useful include HVAC environmental controls, sensing temperatures inside buildings, equipment or machinery, and process monitoring and control.

1 of 27 050400

SUVT Plexiglas

FIGURE 5

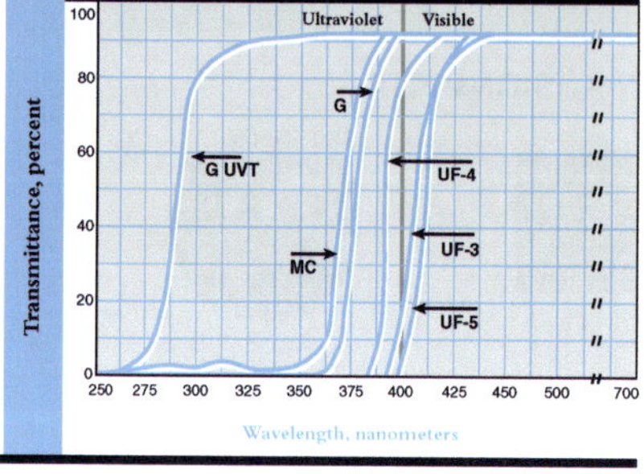

Visible and Ultraviolet Transmittance in Colorless PLEXIGLAS Sheet

occurred in the spectrophotometric curves of these exposed samples between 5 and 10 years outdoors.

Colorless Plexiglas sheet exhibits the same excellent resistance to discoloration when exposed for 20 years or more to constantly lit fluorescent lamps, even when the Plexiglas sheet is only two inches from the lamp.

This unsurpassed UV stability gives Plexiglas sheet superb weatherability and makes it the logical choice among plastic materials for outdoor and artificial lighting applications. Other transparent plastic materials tend to yellow or develop appreciable haze levels when exposed to UV energy.

The spectrophotometric curves in Figure 5 show the light transmission of colorless Plexiglas sheet formulations in the UV bands of the electromagnetic spectrum. Note that regular Plexiglas G and MC sheets block some portions of the UV spectrum, but special grades are needed when virtually all (UF-3/UF-5) or most (UF-4) of the UV wavelengths must be excluded.

PLEXIGLAS ULTRAVIOLET-FILTERING FORMULATIONS

Some special formulations of Plexiglas acrylic plastic have spectrophotometric characteristics that enable them to filter out almost all of the UV energy. These formulations are available in both sheet and molding resins. Plexiglas UV-filtering materials offer protection from structural damage caused by exposure to UV wavelengths shorter than 400 nanometers. However, since these materials do not appreciably filter visible light in the critical lower violet range from 400 to 500 nanometers, they will not completely prevent colors from fading or darkening.

Limiting exposure to light is only one of the steps that can be taken to slow the deterioration of materials. Factors such as temperature, humidity and atmospheric contamination will also affect this process.

Three Plexiglas sheet formulations that can filter UV light are available. Plexiglas UF-4 sheet absorbs most ultraviolet radiation. Plexiglas UF-3 sheet absorbs virtually all ultraviolet radiation and some of the visible light in the violet wavelength region. Plexiglas UF-5 sheet, a new, special formulation in the MC grade for framing applications, filters the maximum of harmful UV radiation. Figure 5 compares the spectrophotometric curves of these three grades with those of standard colorless Plexiglas G and Plexiglas MC sheets. Note that the spectrophotometric characteristics of the three Plexiglas UF sheet formulations are the same, regardless of the thickness of the sheet. Comparatively, the spectrophotometric characteristics of Plexiglas G and Plexiglas MC sheets change with thickness.

Plexiglas UF-3 and MC UF-5 sheets are the most effective of the ultraviolet-filtering formulations of Plexiglas acrylic. Not only do these materials absorb all ultraviolet radiation (390 nanometers and below), but they also absorb part of the relatively harmful visible light in the critical violet (400-450 nanometers) region. Plexiglas UF-3 sheet does absorb some visible light, therefore it has a very faint yellow edge tint. This tint is not objectionable in most applications.

Plexiglas UF-4 sheet may be used for those applications in which the slight yellow tint of the other materials is objectionable, such as unframed display cases. Plexiglas UF-4 does transmit slightly more UV energy than its counterpart, and, as a result, it is somewhat less effective in retarding fading or darkening of colors caused by visible light in the 400 to 500 nanometer range.

Both cast sheet UF formulations are available in thicknesses of 0.118" (3mm) through 0.472" (12mm),

4

LTC3108

ANALOG DEVICES | POWER BY LINEAR

LTC3108

Ultralow Voltage Step-Up Converter and Power Manager

FEATURES

- **Operates from Inputs of 20mV**
- **Complete Energy Harvesting Power Management System**
  - **Selectable V_{OUT} of 2.35V, 3.3V, 4.1V or 5V**
  - **LDO: 2.2V at 3mA**
  - **Logic Controlled Output**
  - **Reserve Energy Output**
- **Power Good Indicator**
- Uses Compact Step-Up Transformers
- Small 12-Lead (3mm × 4mm) DFN or 16-Lead SSOP Packages

APPLICATIONS

- Remote Sensors and Radio Power
- Surplus Heat Energy Harvesting
- HVAC Systems
- Industrial Wireless Sensing
- Automatic Metering
- Building Automation
- Predictive Maintenance

DESCRIPTION

The LTC®3108 is a highly integrated DC/DC converter ideal for harvesting and managing surplus energy from extremely low input voltage sources such as TEGs (thermoelectric generators), thermopiles and small solar cells. The step-up topology operates from input voltages as low as 20mV. The LTC3108 is functionally equivalent to the LTC3108-1 except for its unique fixed V_{OUT} options.

Using a small step-up transformer, the LTC3108 provides a complete power management solution for wireless sensing and data acquisition. The 2.2V LDO powers an external microprocessor, while the main output is programmed to one of four fixed voltages to power a wireless transmitter or sensors. The power good indicator signals that the main output voltage is within regulation. A second output can be enabled by the host. A storage capacitor provides power when the input voltage source is unavailable. Extremely low quiescent current and high efficiency design ensure the fastest possible charge times of the output reservoir capacitor.

The LTC3108 is available in a small, thermally enhanced 12-lead (3mm × 4mm) DFN package and a 16-lead SSOP package.

TYPICAL APPLICATION

Wireless Remote Sensor Application Powered From a Peltier Cell

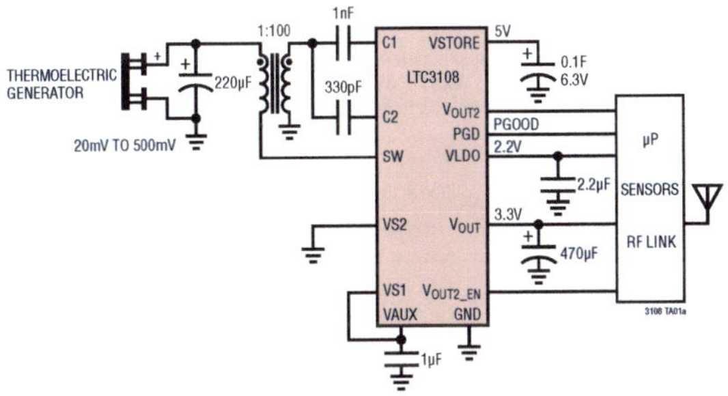

V_{OUT} Charge Time

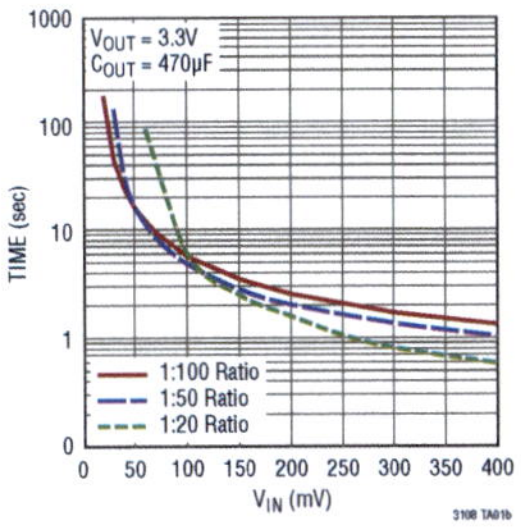

Rev. D

LTC3108

ABSOLUTE MAXIMUM RATINGS (Note 1)

SW Voltage .... –0.3V to 2V
C1 Voltage.. –0.3V to 6V
C2 Voltage (Note 5).. –8V to 8V
V_{OUT2}, V_{OUT2_EN} .... –0.3V to 6V
VAUX.. 15mA into VAUX
VS1, VS2, VAUX, V_{OUT}, PGD .... –0.3V to 6V
VLDO, VSTORE .... –0.3V to 6V
Operating Junction Temperature Range
(Note 2)... –40°C to 125°C
Storage Temperature Range .... –65°C to 125°C

PIN CONFIGURATION

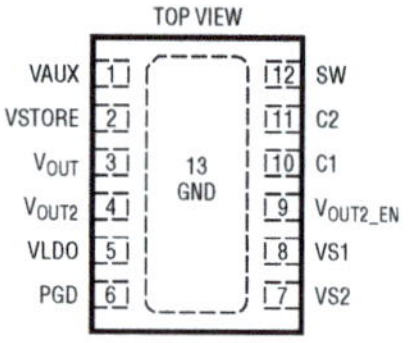

DE PACKAGE
12-LEAD (4mm × 3mm) PLASTIC DFN

T_{JMAX} = 125°C, θ_{JA} = 43°C/W
EXPOSED PAD (PIN 13) IS GND, MUST BE SOLDERED TO PCB (NOTE 4)

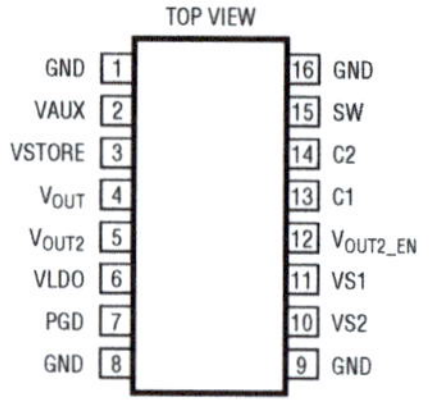

GN PACKAGE
16-LEAD PLASTIC SSOP NARROW

T_{JMAX} = 125°C, θ_{JA} = 110°C/W

ORDER INFORMATION

LEAD FREE FINISH	TAPE AND REEL	PART MARKING*	PACKAGE DESCRIPTION	TEMPERATURE RANGE
LTC3108EDE#PBF	LTC3108EDE#TRPBF	3108	12-Lead (4mm × 3mm) Plastic DFN	–40°C to 125°C
LTC3108IDE#PBF	LTC3108IDE#TRPBF	3108	12-Lead (4mm × 3mm) Plastic DFN	–40°C to 125°C
LTC3108EGN#PBF	LTC3108EGN#TRPBF	3108	16-Lead Plastic SSOP	–40°C to 125°C
LTC3108IGN#PBF	LTC3108IGN#TRPBF	3108	16-Lead Plastic SSOP	–40°C to 125°C

Contact the factory for parts specified with wider operating temperature ranges. *The temperature grade is identified by a label on the shipping container.
Tape and reel specifications. Some packages are available in 500 unit reels through designated sales channels with #TRMPBF suffix.

ELECTRICAL CHARACTERISTICS

The ● denotes the specifications which apply over the full operating junction temperature range, otherwise specifications are for T_A = 25°C (Note 2). VAUX = 5V, unless otherwise noted.

PARAMETER	CONDITIONS		MIN	TYP	MAX	UNITS
Minimum Start-Up Voltage	Using 1:100 Transformer Turns Ratio, VAUX = 0V			20	50	mV
No-Load Input Current	Using 1:100 Transformer Turns Ratio; V_{IN} = 20mV, V_{OUT2_EN} = 0V; All Outputs Charged and in Regulation			3		mA
Input Voltage Range	Using 1:100 Transformer Turns Ratio	●	$V_{STARTUP}$		500	mV

Rev. D

2

For more information www.analog.com

ELECTRICAL CHARACTERISTICS

The ● denotes the specifications which apply over the full operating junction temperature range, otherwise specifications are for $T_A = 25°C$ (Note 2). VAUX = 5V, unless otherwise noted.

PARAMETER	CONDITIONS		MIN	TYP	MAX	UNITS
Output Voltage	VS1 = VS2 = GND	●	2.30	2.350	2.40	V
	VS1 = VAUX, VS2 = GND	●	3.234	3.300	3.366	V
	VS1 = GND, VS2 = VAUX	●	4.018	4.100	4.182	V
	VS1 = VS2 = VAUX	●	4.90	5.000	5.10	V
V_{OUT} Quiescent Current	V_{OUT} = 3.3V, V_{OUT2_EN} = 0V			0.2		μA
VAUX Quiescent Current	No Load, All Outputs Charged			6	9	μA
LDO Output Voltage	0.5mA Load	●	2.134	2.2	2.266	V
LDO Load Regulation	For 0mA to 2mA Load			0.5	1	%
LDO Line Regulation	For VAUX from 2.5V to 5V			0.05	0.2	%
LDO Dropout Voltage	I_{LDO} = 2mA			100	200	mV
LDO Current Limit	V_{LDO} = 0V	●	4	11		mA
V_{OUT} Current Limit	V_{OUT} = 0V	●	2.8	4.5	7.5	mA
VSTORE Current Limit	VSTORE = 0V	●	2.8	4.5	7.5	mA
VAUX Clamp Voltage	Current into VAUX = 5mA	●	5	5.25	5.60	V
VSTORE Leakage Current	VSTORE = 5V			0.1	0.3	μA
V_{OUT2} Leakage Current	V_{OUT2} = 0V, V_{OUT2_EN} = 0V			0.1		μA
VS1, VS2 Threshold Voltage		●	0.4	0.85	1.2	V
VS1, VS2 Input Current	VS1 = VS2 = 5V			0.01	0.1	μA
PGOOD Threshold (Rising)	Measured Relative to the V_{OUT} Voltage			–7.5		%
PGOOD Threshold (Falling)	Measured Relative to the V_{OUT} Voltage			–9		%
PGOOD V_{OL}	Sink Current = 100μA			0.15	0.3	V
PGOOD V_{OH}	Source Current = 0		2.1	2.2	2.3	V
PGOOD Pull-Up Resistance				1		MΩ
V_{OUT2_EN} Threshold Voltage	V_{OUT2_EN} Rising	●	0.4	1	1.3	V
V_{OUT2_EN} Pull-Down Resistance				5		MΩ
V_{OUT2} Turn-On Time				5		μs
V_{OUT2} Turn-Off Time	(Note 3)			0.15		μs
V_{OUT2} Current Limit	V_{OUT} = 3.3V	●	0.15	0.3	0.45	A
V_{OUT2} Current Limit Response Time	(Note 3)			350		ns
V_{OUT2} P-Channel MOSFET On-Resistance	V_{OUT} = 3.3V (Note 3)			1.3		Ω
N-Channel MOSFET On-Resistance	C2 = 5V (Note 3)			0.5		Ω

Note 1: Stresses beyond those listed under Absolute Maximum Ratings may cause permanent damage to the device. Exposure to any Absolute Maximum Rating condition for extended periods may affect device reliability and lifetime.

Note 2: The LTC3108 is tested under pulsed load conditions such that $T_J \approx T_A$. The LTC3108E is guaranteed to meet specifications from 0°C to 85°C junction temperature. Specifications over the –40°C to 125°C operating junction temperature range are assured by design, characterization and correlation with statistical process controls. The LTC3108I is guaranteed over the full –40°C to 125°C operating junction temperature range. Note that the maximum ambient temperature is determined by specific operating conditions in conjunction with board layout, the rated thermal package thermal resistance and other environmental factors. The junction temperature (T_J) is calculated from the ambient temperature (T_A) and power dissipation (P_D) according to the formula: $T_J = T_A + (P_D \bullet \theta_{JA}°C/W)$, where θ_{JA} is the package thermal impedance.

Note 3: Specification is guaranteed by design and not 100% tested in production.

Note 4: Failure to solder the exposed backside of the package to the PC board ground plane will result in a thermal resistance much higher than 43°C/W.

Note 5: The absolute maximum rating is a DC rating. Under certain conditions in the applications shown, the peak AC voltage on the C2 pin may exceed ±8V. This behavior is normal and acceptable because the current into the pin is limited by the impedance of the coupling capacitor.

Rev. D

LTC3108

TYPICAL PERFORMANCE CHARACTERISTICS $T_A = 25°C$, unless otherwise noted.

I_{IN} vs V_{IN}, (V_{OUT} = 0V)

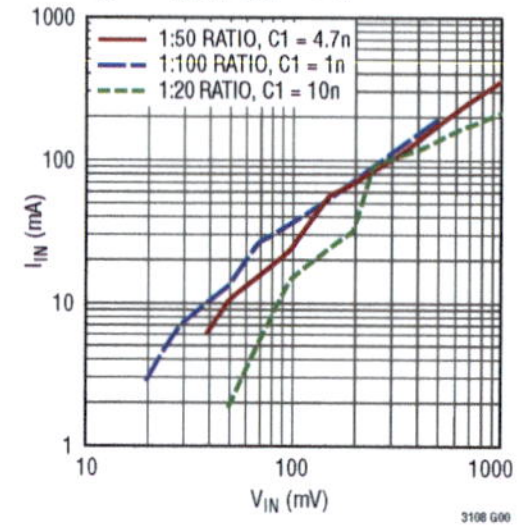

I_{VOUT} and Efficiency vs V_{IN}, 1:20 Ratio Transformer

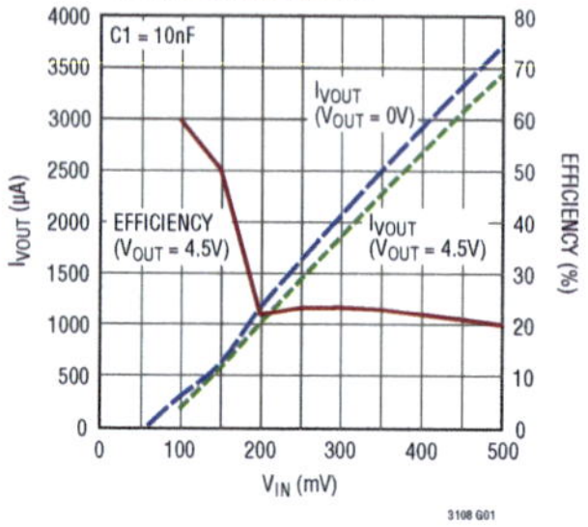

I_{VOUT} and Efficiency vs V_{IN}, 1:100 Ratio Transformer

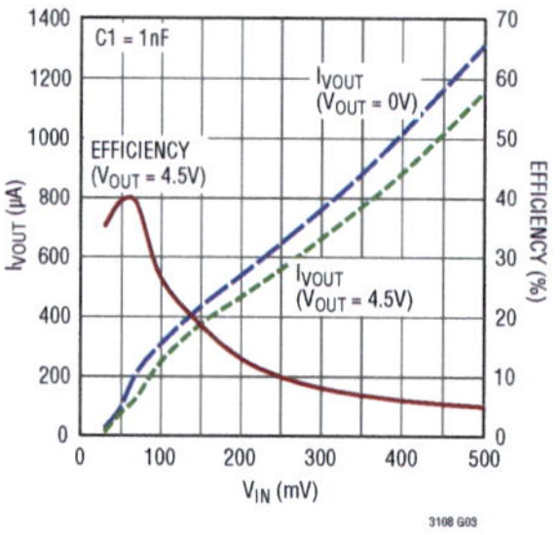

I_{VOUT} and Efficiency vs V_{IN}, 1:50 Ratio Transformer

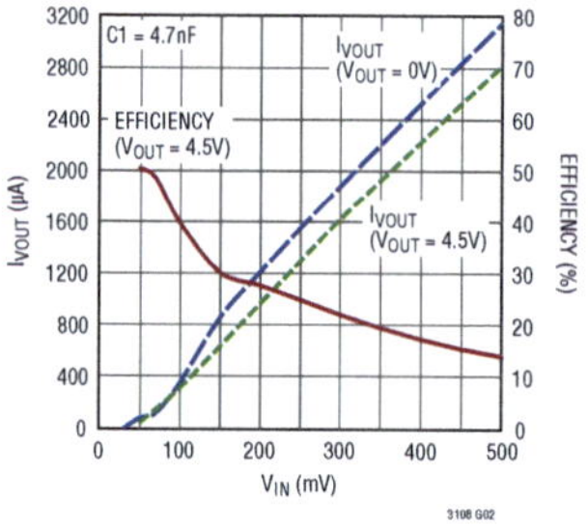

Input Resistance vs V_{IN} (V_{OUT} Charging)

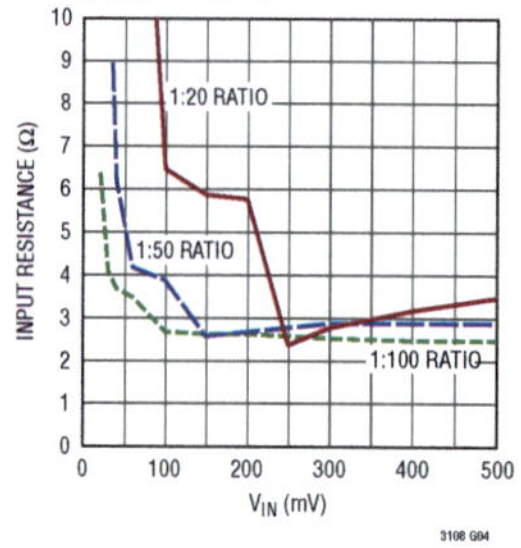

I_{VOUT} vs V_{IN} and Source Resistance, 1:20 Ratio

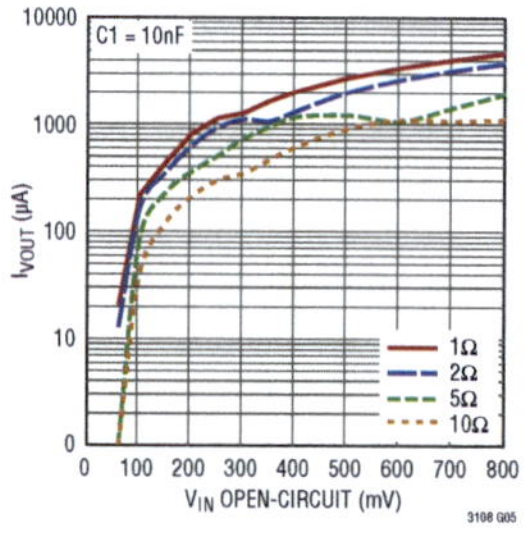

Rev. D

4

For more information www.analog.com

LTC3108

TYPICAL PERFORMANCE CHARACTERISTICS T_A = 25°C, unless otherwise noted.

I_{VOUT} vs V_{IN} and Source Resistance, 1:50 Ratio

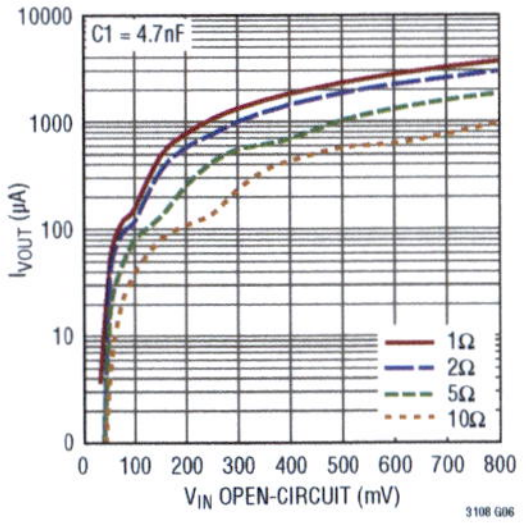

I_{VOUT} vs V_{IN} and Source Resistance, 1:100 Ratio

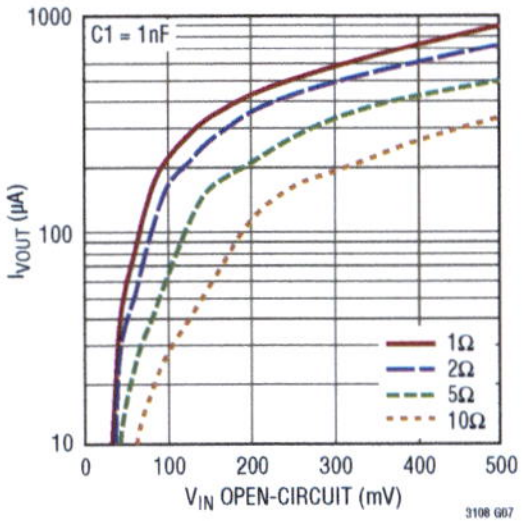

I_{VOUT} vs dT and TEG Size, 1:100 Ratio

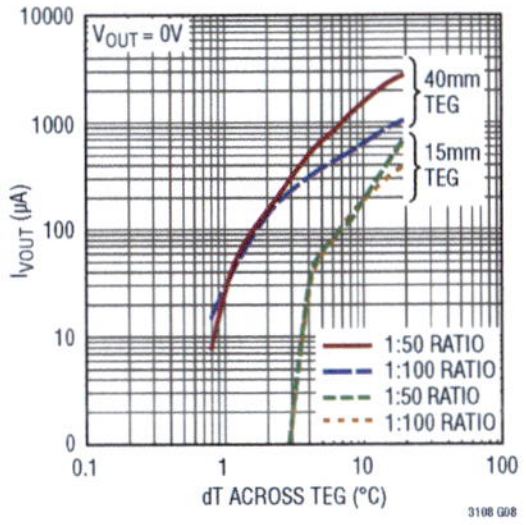

Resonant Switching Waveforms

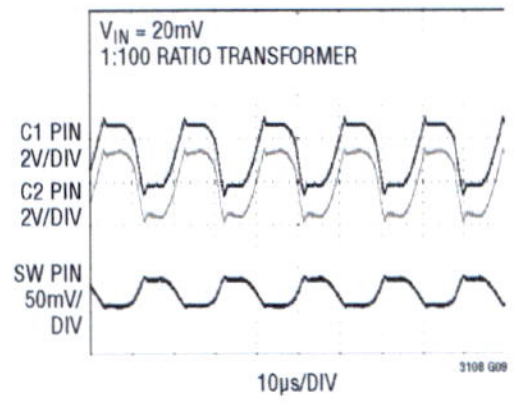

LDO Load Regulation

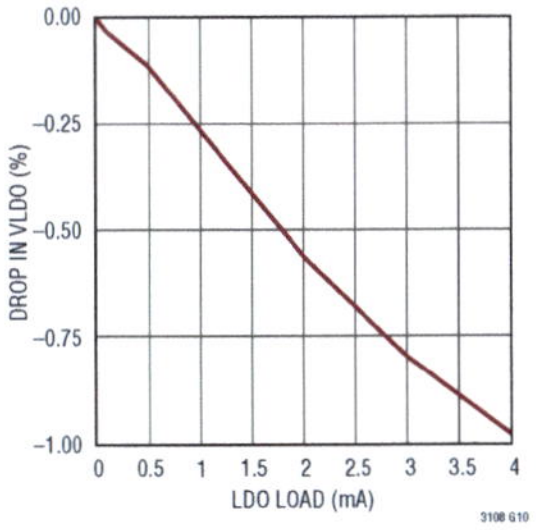

LDO Dropout Voltage

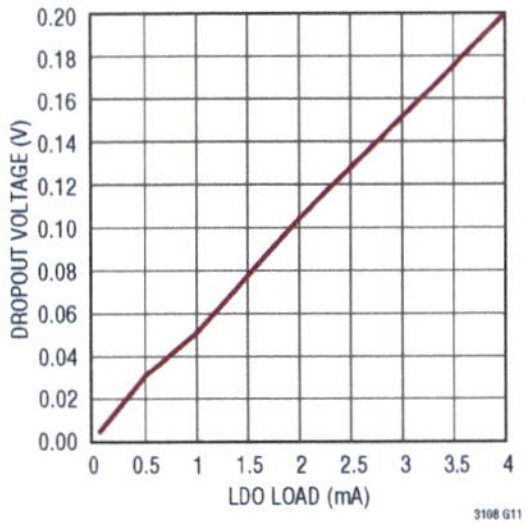

Rev. D

LTC3108

TYPICAL PERFORMANCE CHARACTERISTICS T_A = 25°C, unless otherwise noted.

Start-Up Voltage Sequencing

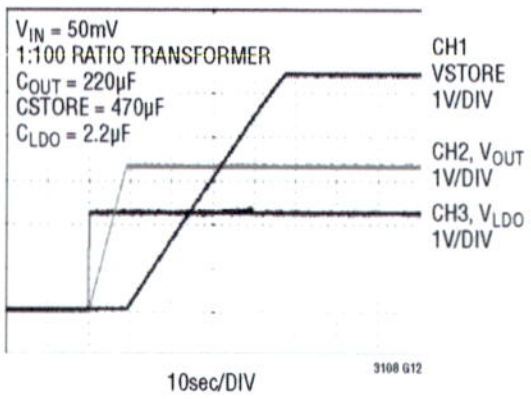

V_{OUT} and PGD Response During a Step Load

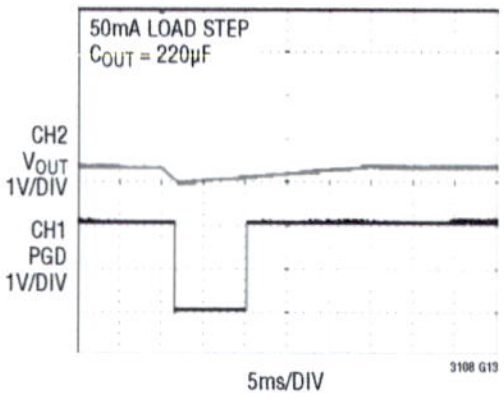

V_{OUT} Ripple

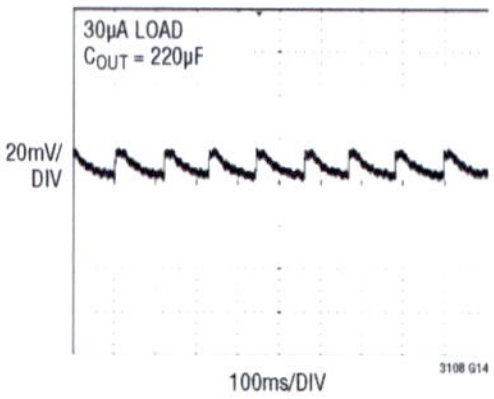

LDO Step Load Response

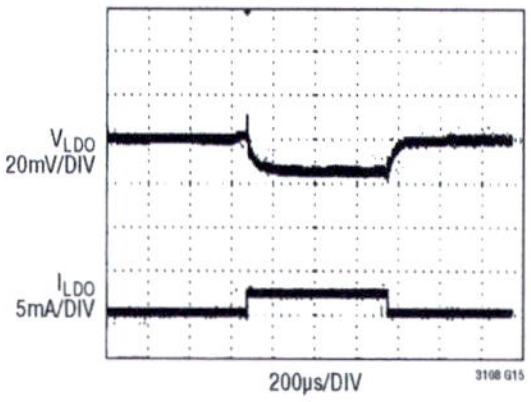

0mA TO 3mA LOAD STEP
C_{LDO} = 2.2µF

Enable Input and V_{OUT2}

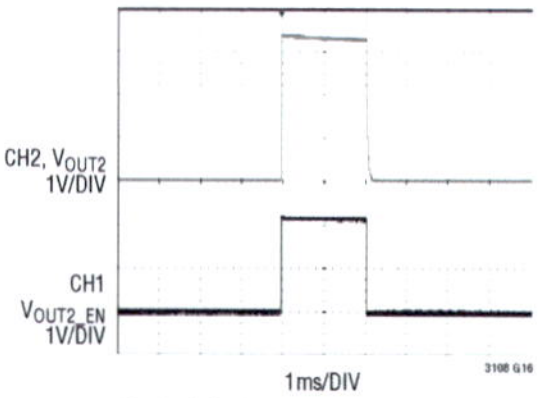

10mA LOAD ON V_{OUT2}
C_{OUT} = 220µF

Running on Storage Capacitor

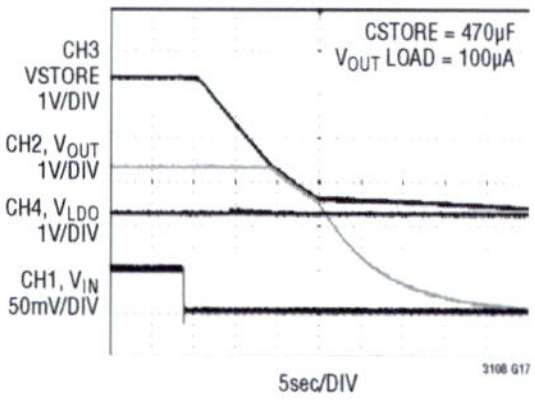

PIN FUNCTIONS (DFN/SSOP)

VAUX (Pin 1/Pin 2): Output of the Internal Rectifier Circuit and V_{CC} for the IC. Bypass VAUX with at least 1µF of capacitance. An active shunt regulator clamps VAUX to 5.25V (typical).

VSTORE (Pin 2/Pin 3): Output for the Storage Capacitor or Battery. A large capacitor may be connected from this pin to GND for powering the system in the event the input voltage is lost. It will be charged up to the maximum VAUX clamp voltage. If not used, this pin should be left open or tied to VAUX.

V_{OUT} (Pin 3/Pin 4): Main Output of the Converter. The voltage at this pin is regulated to the voltage selected by VS1 and VS2 (see Table 1). Connect this pin to an energy storage capacitor or to a rechargeable battery.

V_{OUT2} (Pin 4/Pin 5): Switched Output of the Converter. Connect this pin to a switched load. This output is open until V_{OUT2_EN} is driven high, then it is connected to V_{OUT} through a 1.3Ω P-channel switch. If not used, this pin should be left open or tied to V_{OUT}. The peak current in this output is limited to 0.3A typical.

VLDO (Pin 5/Pin 6): Output of the 2.2V LDO. Connect a 2.2µF or larger ceramic capacitor from this pin to GND. If not used, this pin should be tied to VAUX.

PGD (Pin 6/Pin 7): Power Good Output. When V_{OUT} is within 7.5% of its programmed value, PGD will be pulled up to VLDO through a 1MΩ resistor. If V_{OUT} drops 9% below its programmed value PGD will go low. This pin can sink up to 100µA.

VS2 (Pin 7/Pin 10): V_{OUT} Select Pin 2. Connect this pin to ground or VAUX to program the output voltage (see Table 1).

VS1 (Pin 8/Pin 11): V_{OUT} Select Pin 1. Connect this pin to ground or VAUX to program the output voltage (see Table 1).

V_{OUT2_EN} (Pin 9/Pin 12): Enable Input for V_{OUT2}. V_{OUT2} will be enabled when this pin is driven high. There is an internal 5M pull-down resistor on this pin. If not used, this pin can be left open or grounded.

C1 (Pin 10/Pin 13): Input to the Charge Pump and Rectifier Circuit. Connect a capacitor from this pin to the secondary winding of the step-up transformer.

C2 (Pin 11/Pin 14): Input to the N-Channel Gate Drive Circuit. Connect a capacitor from this pin to the secondary winding of the step-up transformer.

SW (Pin 12/Pin 15): Drain of the Internal N-Channel Switch. Connect this pin to the primary winding of the transformer.

GND (Pins 1, 8, 9, 16) SSOP Only: Ground

GND (Exposed Pad Pin 13) DFN Only: Ground. The DFN exposed pad must be soldered to the PCB ground plane. It serves as the ground connection, and as a means of conducting heat away from the die.

Table 1. Regulated Voltage Using Pins VS1 and VS2

VS2	VS1	V_{OUT}
GND	GND	2.35V
GND	VAUX	3.3V
VAUX	GND	4.1V
VAUX	VAUX	5V

BLOCK DIAGRAM

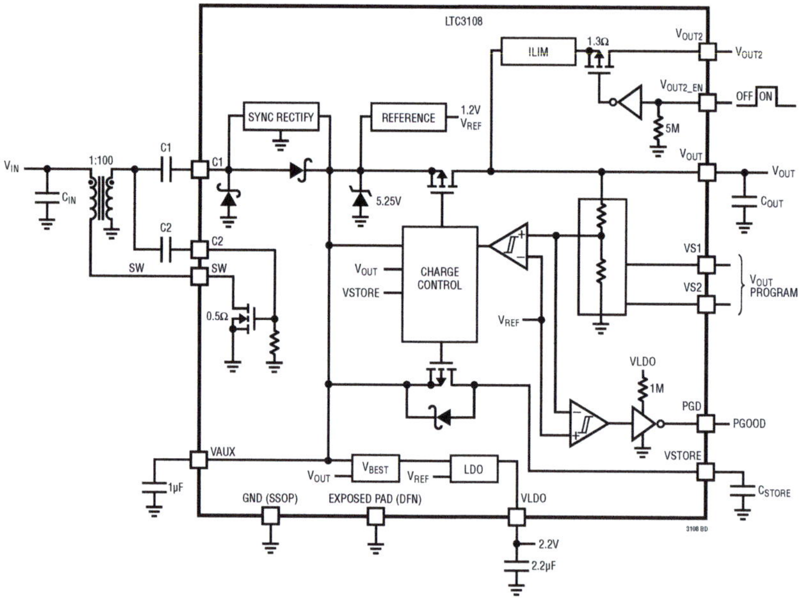

OPERATION (Refer to the Block Diagram)

The LTC3108 is designed to use a small external step-up transformer to create an ultralow input voltage step-up DC/DC converter and power manager. It is ideally suited for low power wireless sensors and other applications in which surplus energy harvesting is used to generate system power because traditional battery power is inconvenient or impractical.

The LTC3108 is designed to manage the charging and regulation of multiple outputs in a system in which the average power draw is very low, but there may be periodic pulses of higher load current required. This is typical of wireless sensor applications, where the quiescent power draw is extremely low most of the time, except for transmit bursts when circuitry is powered up to make measurements and transmit data.

The LTC3108 can also be used to trickle charge a standard capacitor, supercapacitor or rechargeable battery, using energy harvested from a Peltier or photovoltaic cell.

OPERATION

Oscillator

The LTC3108 utilizes a MOSFET switch to form a resonant step-up oscillator using an external step-up transformer and a small coupling capacitor. This allows it to boost input voltages as low as 20mV high enough to provide multiple regulated output voltages for powering other circuits. The frequency of oscillation is determined by the inductance of the transformer secondary winding and is typically in the range of 10kHz to 100kHz. For input voltages as low as 20mV, a primary-secondary turns ratio of about 1:100 is recommended. For higher input voltages, this ratio can be lower. See the Applications Information section for more information on selecting the transformer.

Charge Pump and Rectifier

The AC voltage produced on the secondary winding of the transformer is boosted and rectified using an external charge pump capacitor (from the secondary winding to pin C1) and the rectifiers internal to the LTC3108. The rectifier circuit feeds current into the VAUX pin, providing charge to the external VAUX capacitor and the other outputs.

VAUX

The active circuits within the LTC3108 are powered from VAUX, which should be bypassed with a 1µF capacitor. Larger capacitor values are recommended when using turns ratios of 1:50 or 1:20 (refer to the Typical Application examples). Once VAUX exceeds 2.5V, the main V_{OUT} is allowed to start charging.

An internal shunt regulator limits the maximum voltage on VAUX to 5.25V typical. It shunts to GND any excess current into VAUX when there is no load on the converter or the input source is generating more power than is required by the load.

Voltage Reference

The LTC3108 includes a precision, micropower reference, for accurate regulated output voltages. This reference becomes active as soon as VAUX exceeds 2V.

Synchronous Rectifiers

Once VAUX exceeds 2V, synchronous rectifiers in parallel with each of the internal diodes take over the job of rectifying the input voltage, improving efficiency.

Low Dropout Linear Regulator (LDO)

The LTC3108 includes a low current LDO to provide a regulated 2.2V output for powering low power processors or other low power ICs. The LDO is powered by the higher of VAUX or V_{OUT}. This enables it to become active as soon as VAUX has charged to 2.3V, while the V_{OUT} storage capacitor is still charging. In the event of a step load on the LDO output, current can come from the main V_{OUT} capacitor if VAUX drops below V_{OUT}. The LDO requires a 2.2µF ceramic capacitor for stability. Larger capacitor values can be used without limitation, but will increase the time it takes for all the outputs to charge up. The LDO output is current limited to 4mA minimum.

V_{OUT}

The main output voltage on V_{OUT} is charged from the VAUX supply, and is user programmed to one of four regulated voltages using the voltage select pins VS1 and VS2, according to Table 2. Although the logic threshold voltage for VS1 and VS2 is 0.85V typical, it is recommended that they be tied to ground or VAUX.

Table 2. Regulated Voltage Using Pins VS1 and VS2

VS2	VS1	V_{OUT}
GND	GND	2.35V
GND	VAUX	3.3V
VAUX	GND	4.1V
VAUX	VAUX	5V

When the output voltage drops slightly below the regulated value, the charging current will be enabled as long as VAUX is greater than 2.5V. Once V_{OUT} has reached the proper value, the charging current is turned off.

The internal programmable resistor divider sets V_{OUT}, eliminating the need for very high value external resistors that are susceptible to board leakage.

OPERATION

In a typical application, a storage capacitor (typically a few hundred microfarads) is connected to V_{OUT}. As soon as VAUX exceeds 2.5V, the V_{OUT} capacitor will be allowed to charge up to its regulated voltage. The current available to charge the capacitor will depend on the input voltage and transformer turns ratio, but is limited to about 4.5mA typical.

PGOOD

A power good comparator monitors the V_{OUT} voltage. The PGD pin is an open-drain output with a weak pull-up (1MΩ) to the LDO voltage. Once V_{OUT} has charged to within 7.5% of its regulated voltage, the PGD output will go high. If V_{OUT} drops more than 9% from its regulated voltage, PGD will go low. The PGD output is designed to drive a microprocessor or other chip I/O and is not intended to drive a higher current load such as an LED. Pulling PGD up externally to a voltage greater than VLDO will cause a small current to be sourced into VLDO. PGD can be pulled low in a wire-OR configuration with other circuitry.

V_{OUT2}

V_{OUT2} is an output that can be turned on and off by the host, using the V_{OUT2_EN} pin. When enabled, V_{OUT2} is connected to V_{OUT} through a 1.3Ω P-channel MOSFET switch. This output, controlled by a host processor, can be used to power external circuits such as sensors and amplifiers, that do not have a low power sleep or shutdown capability. V_{OUT2} can be used to power these circuits only when they are needed.

Minimizing the amount of decoupling capacitance on V_{OUT2} will allow it to be switched on and off faster, allowing shorter burst times and, therefore, smaller duty cycles in pulsed applications such as a wireless sensor/transmitter. A small V_{OUT2} capacitor will also minimize the energy that will be wasted in charging the capacitor every time V_{OUT2} is enabled.

V_{OUT2} has a soft-start time of about 5μs to limit capacitor charging current and minimize glitching of the main output when V_{OUT2} is enabled. It also has a current limiting circuit that limits the peak current to 0.3A typical.

The V_{OUT2} enable input has a typical threshold of 1V with 100mV of hysteresis, making it logic-compatible. If V_{OUT2_EN} (which has an internal pull-down resistor) is low, V_{OUT2} will be off. Driving V_{OUT2_EN} high will turn on the V_{OUT2} output.

Note that while V_{OUT2_EN} is high, the current limiting circuitry for V_{OUT2} draws an extra 8μA of quiescent current from V_{OUT}. This added current draw has a negligible effect on the application and capacitor sizing, since the load on the V_{OUT2} output, when enabled, is likely to be orders of magnitude higher than 8μA.

VSTORE

The VSTORE output can be used to charge a large storage capacitor or rechargeable battery after V_{OUT} has reached regulation. Once V_{OUT} has reached regulation, the VSTORE output will be allowed to charge up to the VAUX voltage. The storage element on VSTORE can be used to power the system in the event that the input source is lost, or is unable to provide the current demanded by the V_{OUT}, V_{OUT2} and LDO outputs. If VAUX drops below VSTORE, the LTC3108 will automatically draw current from the storage element. Note that it may take a long time to charge a large capacitor, depending on the input energy available and the loading on V_{OUT} and VLDO.

Since the maximum current from VSTORE is limited to a few milliamps, it can safely be used to trickle-charge NiCd or NiMH rechargeable batteries for energy storage when the input voltage is lost. Note that the VSTORE capacitor cannot supply large pulse currents to V_{OUT}. Any pulse load on V_{OUT} must be handled by the V_{OUT} capacitor.

Short-Circuit Protection

All outputs of the LTC3108 are current limited to protect against short-circuits to ground.

Output Voltage Sequencing

A timing diagram showing the typical charging and voltage sequencing of the outputs is shown in Figure 1. Note: time not to scale.

OPERATION

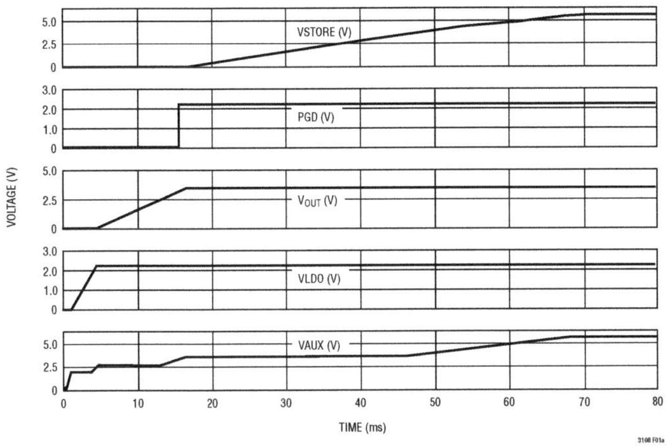

Figure 1. Output Voltage Sequencing with V_{OUT} Programmed for 3.3V (Time Not to Scale)

APPLICATIONS INFORMATION

Introduction

The LTC3108 is designed to gather energy from very low input voltage sources and convert it to usable output voltages to power microprocessors, wireless transmitters and analog sensors. Such applications typically require much more peak power, and at higher voltages, than the input voltage source can produce. The LTC3108 is designed to accumulate and manage energy over a long period of time to enable short power bursts for acquiring and transmitting data. The bursts must occur at a low enough duty cycle such that the total output energy during the burst does not exceed the average source power integrated over the accumulation time between bursts. For many applications, this time between bursts could be seconds, minutes or hours.

The PGD signal can be used to enable a sleeping microprocessor or other circuitry when V_{OUT} reaches regulation, indicating that enough energy is available for a burst.

Input Voltage Sources

The LTC3108 can operate from a number of low input voltage sources, such as Peltier cells, photovoltaic cells or thermopile generators. The minimum input voltage required for a given application will depend on the transformer turns ratio, the load power required, and the internal DC resistance (ESR) of the voltage source. Lower ESR will allow the use of lower input voltages, and provide higher output power capability.

Refer to the I_{IN} vs V_{IN} curves in the Typical Performance Characteristics section to see what input current is required from the source for a given input voltage.

For a given transformer turns ratio, there is a maximum recommended input voltage to avoid excessively high secondary voltages and power dissipation in the shunt regulator. It is recommended that the maximum input voltage times the turns ratio be less than 50.

Note that a low ESR bulk decoupling capacitor will usually be required across the input source to prevent large voltage droop and ripple caused by the source's ESR and the peak primary switching current (which can reach hundreds of milliamps). The time constant of the filter capacitor and the ESR of the voltage source should be much longer than the period of the resonant switching frequency.

Peltier Cell (Thermoelectric Generator)

A Peltier cell (also known as a thermoelectric cooler) is made up of a large number of series-connected P-N junctions, sandwiched between two parallel ceramic plates. Although Peltier cells are often used as coolers by applying a DC voltage to their inputs, they will also generate a DC output voltage, using the Seebeck effect, when the two plates are at different temperatures. The polarity of the output voltage will depend on the polarity of the temperature differential between the plates. The magnitude of the output voltage is proportional to the magnitude of the

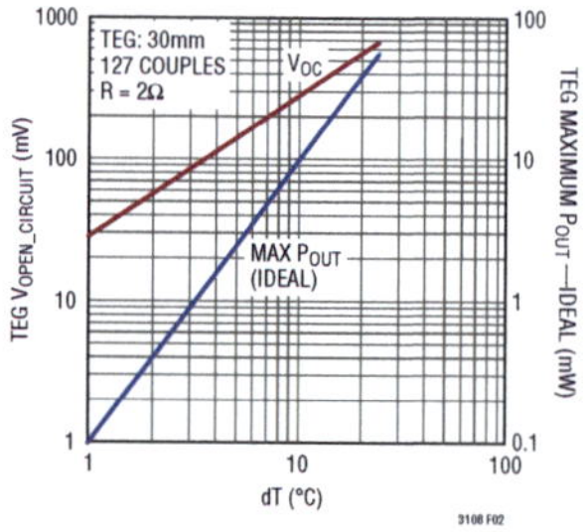

Figure 2. Typical Performance of a Peltier Cell Acting as a Thermoelectric Generator

APPLICATIONS INFORMATION

temperature differential between the plates. When used in this manner, a Peltier cell is referred to as a thermoelectric generator (TEG).

The low voltage capability of the LTC3108 design allows it to operate from a TEG with temperature differentials as low as 1°C, making it ideal for harvesting energy in applications in which a temperature difference exists between two surfaces or between a surface and the ambient temperature. The internal resistance (ESR) of most cells is in the range of 1Ω to 5Ω, allowing for reasonable power transfer. The curves in Figure 2 show the open-circuit output voltage and maximum power transfer for a typical Peltier cell (with an ESR of 2Ω) over a 20°C range of temperature differential.

TEG Load Matching

The LTC3108 was designed to present a minimum input resistance (load) in the range of 2Ω to 10Ω, depending on input voltage and transformer turns ratio (as shown in the Typical Performance Characteristics curves). For a given turns ratio, as the input voltage drops, the input resistance increases. This feature allows the LTC3108 to optimize power transfer from sources with a few ohms of source resistance, such as a typical TEG. Note that a lower source resistance will always provide more output current capability by providing a higher input voltage under load.

Peltier Cell (TEG) Suppliers

Peltier cells are available in a wide range of sizes and power capabilities, from less than 10mm square to over 50mm square. They are typically 2mm to 5mm in height. A list of Peltier cell manufacturers is given in Table 3.

Table 3. Peltier Cell Manufacturers

CUI, Inc. www.cui.com (Distributor)
Fujitaka www.fujitaka.com/pub/peltier/english/thermoelectric_power.html
Ferrotec www.ferrotec.com/products/thermal/modules
Kryotherm www.kryothermusa.com
Laird Technologies www.lairdtech.com
Marlow Industries www.marlow.com
Micropelt www.micropelt.com
Nextreme www.nextreme.com
TE Technology www.tetech.com/Peltier-Thermoelectric-Cooler-Modules.html
Tellurex www.tellurex.com

Table 4. Recommended TEG Part Numbers by Size

MANUFACTURER	15mm × 15mm	20mm × 20mm	30mm × 30mm	40mm × 40mm
CUI Inc. (Distributor)	CP60133	CP60233	CP60333	CP85438
Ferrotec	9501/031/030 B	9501/071/040 B	9500/097/090 B	9500/127/100 B
Fujitaka	FPH13106NC	FPH17106NC	FPH17108AC	FPH112708AC
Kryotherm			TGM-127-1.0-0.8	LCB-127-1.4-1.15
Laird Technology			PT6.7.F2.3030.W6	PT8.12.F2.4040.TA.W6
Marlow Industries		RC3-8-01	RC6-6-01	RC12-8-01LS
Tellurex	C2-15-0405	C2-20-0409	C2-30-1505	C2-40-1509
TE Technology	TE-31-1.0-1.3	TE-31-1.4-1.15	TE-71-1.4-1.15	TE-127-1.4-1.05

APPLICATIONS INFORMATION

Thermopile Generator

Thermopile generators (also called powerpile generators) are made up of a number of series-connected thermocouples enclosed in a metal tube. They are commonly used in gas burner applications to generate a DC output of hundreds of millivolts when exposed to the high temperature of a flame. Typical examples are the Honeywell CQ200 and Q313. These devices have an internal series resistance of less than 3Ω, and can generate as much as 750mV open-circuit at their highest rated temperature. For applications in which the temperature rise is too high for a solid-state thermoelectric device, a thermopile can be used as an energy source to power the LTC3108. Because of the higher output voltages possible with a thermopile generator, a lower transformer turns ratio can be used (typically 1:20, depending on the application).

Photovoltaic Cell

The LTC3108 converter can also operate from a single photovoltaic cell (also known as a PV or solar cell) at light levels too low for other low input voltage boost converters to operate. However, many variables will affect the performance in these applications. Light levels can vary over several orders of magnitude and depend on lighting conditions (the type of lighting and indoor versus outdoor). Different types of light (sunlight, incandescent, fluorescent) also have different color spectra, and will produce different output power levels depending on which type of photovoltaic cell is being used (monocrystalline, polycrystalline or thin-film). Therefore, the photovoltaic cell must be chosen for the type and amount of light available. Note that the short-circuit output current from the cell must be at least a few milliamps in order to power the LTC3108 converter

Non-Boost Applications

The LTC3108 can also be used as an energy harvester and power manager for input sources that do not require boosting. In these applications the step-up transformer can be eliminated.

Any source whose peak voltage exceeds 2.5V AC or 5V DC can be connected to the C1 input through a current-limiting resistor where it will be rectified/peak detected. In these applications the C2 and SW pins are not used and can be grounded or left open.

Examples of such input sources would be piezoelectric transducers, vibration energy harvesters, low current generators, a stack of low current solar cells or a 60Hz AC input.

A series resistance of at least 100Ω/V should be used to limit the maximum current into the VAUX shunt regulator.

COMPONENT SELECTION

Step-Up Transformer

The step-up transformer turns ratio will determine how low the input voltage can be for the converter to start. Using a 1:100 ratio can yield start-up voltages as low as 20mV. Other factors that affect performance are the DC resistance of the transformer windings and the inductance of the windings. Higher DC resistance will result in lower efficiency. The secondary winding inductance will determine the resonant frequency of the oscillator, according to the following formula.

$$\text{Frequency} = \frac{1}{2 \bullet \pi \bullet \sqrt{L(sec) \bullet C}}\ \text{Hz}$$

Where L is the inductance of the transformer secondary winding and C is the load capacitance on the secondary winding. This is comprised of the input capacitance at pin C2, typically 30pF, in parallel with the transformer secondary winding's shunt capacitance. The recommended resonant frequency is in the range of 10kHz to 100kHz. See Table 5 for some recommended transformers.

Table 5. Recommended Transformers

VENDOR	PART NUMBER
Coilcraft www.coilcraft.com	LPR6235-752SML (1:100 Ratio) LPR6235-253PML (1:20 Ratio) LPR6235-123QML (1:50 Ratio)
Würth www.we-online	74488540070 (1:100 Ratio) 74488540120 (1:50 Ratio) 74488540250 (1:20 Ratio)

APPLICATIONS INFORMATION

C1 Capacitor

The charge pump capacitor that is connected from the transformer's secondary winding to the C1 pin has an effect on converter input resistance and maximum output current capability. Generally, a minimum value of 1nF is recommended when operating from very low input voltages using a transformer with a ratio of 1:100. Too large a capacitor value can compromise performance when operating at low input voltage or with high resistance sources. For higher input voltages and lower turns ratios, the value of the C1 capacitor can be increased for higher output current capability. Refer to the Typical Applications schematic examples for the recommended value for a given turns ratio.

Squegging

Certain types of oscillators, including transformer-coupled oscillators such as the resonant oscillator of the LTC3108, can exhibit a phenomenon called squegging. This term refers to a condition that can occur which blocks or stops the oscillation for a period of time much longer than the period of oscillation, resulting in bursts of oscillation. An example of this is the blocking oscillator, which is designed to squegg to produce bursts of oscillation. Squegging is also encountered in RF oscillators and regenerative receivers.

In the case of the LTC3108, squegging can occur when a charge builds up on the C2 gate coupling capacitor, such that the DC bias point shifts and oscillation is extinguished for a certain period of time, until the charge on the capacitor bleeds off, allowing oscillation to resume. It is difficult to predict when and if squegging will occur in a given application. While squegging is not harmful, it reduces the average output current capability of the LTC3108.

Squegging can easily be avoided by the addition of a bleeder resistor in parallel with the coupling capacitor on the C2 pin. Resistor values in the range of 100k to 1MΩ are sufficient to eliminate squegging without having any negative impact on performance. For the 330pF capacitor used for C2 in most applications, a 499k bleeder resistor is recommended. See the Typical Applications schematics for an example.

Using External Charge Pump Rectifiers

The synchronous charge pump rectifiers in the LTC3108 (connected to the C1 pin) are optimized for operation from very low input voltage sources, using typical transformer step-up ratios between 1:100 and 1:50, and typical C1 charge pump capacitor values less than 10nF.

Operation from higher input voltage sources (typically 250mV or greater, under load), allows the use of lower transformer step-up ratios (such as 1:20 and 1:10) and larger C1 capacitor values to provide higher output current capability from the LTC3108. However, due to the resulting increase in rectifier currents and resonant oscillator frequency in these applications, the use of external charge pump rectifiers is recommended for optimal performance.

In applications where the step-up ratio is 1:20 or less, and the C1 capacitor is 10nF or greater, the C1 pin should be grounded and two external rectifiers (such as 1N4148 or 1N914 diodes) should be used. These are available as dual diodes in a single package. Avoid the use of Schottky rectifiers, as their lower forward voltage drop increases the minimum start-up voltage. See the Typical Applications schematics for an example.

V_{OUT} and VSTORE Capacitor

For pulsed load applications, the V_{OUT} capacitor should be sized to provide the necessary current when the load is pulsed on. The capacitor value required will be dictated by the load current, the duration of the load pulse, and the amount of voltage droop the circuit can tolerate. The capacitor must be rated for whatever voltage has been selected for V_{OUT} by VS1 and VS2.

$$C_{OUT}(\mu F) \geq \frac{I_{LOAD}(mA) \cdot t_{PULSE}(ms)}{V_{OUT}(V)}$$

Note that there must be enough energy available from the input voltage source for V_{OUT} to recharge the capacitor during the interval between load pulses (to be discussed in the next example). Reducing the duty cycle of the load pulse will allow operation with less input energy.

The VSTORE capacitor may be of very large value (thousands of microfarads or even Farads), to provide holdup

APPLICATIONS INFORMATION

at times when the input power may be lost. Note that this capacitor can charge all the way to 5.25V (regardless of the settings for V_{OUT}), so ensure that the holdup capacitor has a working voltage rating of at least 5.5V at the temperature for which it will be used. The VSTORE capacitor can be sized using the following:

$$C_{STORE} \geq \frac{[6\mu A + I_Q + I_{LDO} + (I_{BURST} \bullet t \bullet f)] \bullet TSTORE}{5.25 - V_{OUT}}$$

Where 6μA is the quiescent current of the LTC3108, I_Q is the load on V_{OUT} in between bursts, I_{LDO} is the load on the LDO between bursts, I_{BURST} is the total load during the burst, t is the duration of the burst, f is the frequency of the bursts, TSTORE is the storage time required and V_{OUT} is the output voltage required. Note that for a programmed output voltage of 5V, the VSTORE capacitor cannot provide any beneficial storage time.

To minimize losses and capacitor charge time, all capacitors used for V_{OUT} and VSTORE should be low leakage. See Table 6 for recommended storage capacitors.

Storage capacitors requiring voltage balancing are not recommended due to the current draw of the balancing resistors.

Table 6. Recommended Storage Capacitors

VENDOR	PART NUMBER/SERIES
AVX www.avx.com	BestCap Series TAJ and TPS Series Tantalum
Cap-XX www.cap-xx.com	GZ Series
Cooper/Bussmann www.bussmann.com/3/PowerStor.html	KR Series P Series
Vishay/Sprague www.vishay.com/capacitors	Tantamount 592D 595D Tantalum 150CRZ/153CRV Aluminum 013 RLC (Low Leakage)

PCB Layout Guidelines

Due to the rather low switching frequency of the resonant converter and the low power levels involved, PCB layout is not as critical as with many other DC/DC converters. There are, however, a number of things to consider.

Due to the very low input voltage the circuit may operate from, the connections to V_{IN}, the primary of the transformer and the SW and GND pins of the LTC3108 should be designed to minimize voltage drop from stray resistance and able to carry currents as high as 500mA. Any small voltage drop in the primary winding conduction path will lower efficiency and increase capacitor charge time.

Also, due to the low charge currents available at the outputs of the LTC3108, any sources of leakage current on the output voltage pins must be minimized. An example board layout is shown in Figure 3.

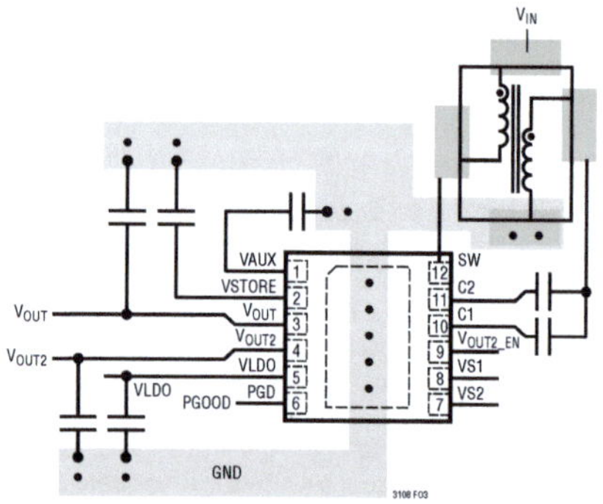

• VIAS TO GROUND PLANE

Figure 3. Example Component Placement for Two-Layer PC Board (DFN Package)

Design Example 1

This design example will explain how to calculate the necessary storage capacitor value for V_{OUT} in pulsed load applications, such as a wireless sensor/transmitter. In these types of applications, the load is very small for a majority of the time (while the circuitry is in a low power sleep state), with bursts of load current occurring periodically during a transmit burst. The storage capacitor on V_{OUT} supports the load during the transmit burst, and the long sleep time between bursts allows the LTC3108 to recharge the capacitor. A method for calculating the

APPLICATIONS INFORMATION

maximum rate at which the load pulses can occur for a given output current from the LTC3108 will also be shown.

In this example, V_{OUT} is set to 3.3V, and the maximum allowed voltage droop during a transmit burst is 10%, or 0.33V. The duration of a transmit burst is 1ms, with a total average current requirement of 40mA during the burst. Given these factors, the minimum required capacitance on V_{OUT} is:

$$C_{OUT}(\mu F) \geq \frac{40mA \bullet 1ms}{0.33V} = 121\mu F$$

Note that this equation neglects the effect of capacitor ESR on output voltage droop. For most ceramic or low ESR tantalum capacitors, the ESR will have a negligible effect at these load currents.

A standard value of 150µF or larger could be used for C_{OUT} in this case. Note that the load current is the total current draw on V_{OUT}, V_{OUT2} and VLDO, since the current for all of these outputs must come from V_{OUT} during a burst. Current contribution from the holdup capacitor on VSTORE is not considered, since it may not be able to recharge between bursts. Also, it is assumed that the charge current from the LTC3108 is negligible compared to the magnitude of the load current during the burst.

To calculate the maximum rate at which load bursts can occur, determine how much charge current is available from the LTC3108 V_{OUT} pin given the input voltage source being used. This number is best found empirically, since there are many factors affecting the efficiency of the converter. Also determine what the total load current is on V_{OUT} during the sleep state (between bursts). Note that this must include any losses, such as storage capacitor leakage.

Assume, for instance, that the charge current from the LTC3108 is 50µA and the total current drawn on V_{OUT} in the sleep state is 17µA, including capacitor leakage. In addition, use the value of 150µF for the V_{OUT} capacitor. The maximum transmit rate (neglecting the duration of the transmit burst, which is typically very short) is then given by:

$$t = \frac{150\mu F \bullet 0.33V}{(50\mu A - 17\mu A)} = 1.5\text{sec or } f_{MAX} = 0.666Hz$$

Therefore, in this application example, the circuit can support a 1ms transmit burst every 1.5 seconds.

It can be determined that for systems that only need to transmit every few seconds (or minutes or hours), the average charge current required is extremely small, as long as the sleep current is low. Even if the available charge current in the example above was only 10µA and the sleep current was only 5µA, it could still transmit a burst every ten seconds.

The following formula enables the user to calculate the time it will take to charge the LDO output capacitor and the V_{OUT} capacitor the first time, from 0V. Here again, the charge current available from the LTC3108 must be known. For this calculation, it is assumed that the LDO output capacitor is 2.2µF.

$$t_{LDO} = \frac{2.2V \bullet 2.2\mu F}{I_{CHG} - I_{LDO}}$$

If there were 50µA of charge current available and a 5µA load on the LDO (when the processor is sleeping), the time for the LDO to reach regulation would be 107ms.

If V_{OUT} were programmed to 3.3V and the V_{OUT} capacitor was 150µF, the time for V_{OUT} to reach regulation would be:

$$t_{VOUT} = \frac{3.3V \bullet 150\mu F}{I_{CHG} - I_{VOUT} - I_{LDO}} + t_{LDO}$$

If there were 50µA of charge current available and 5µA of load on V_{OUT}, the time for V_{OUT} to reach regulation after the initial application of power would be 12.5 seconds.

Design Example 2

In many pulsed load applications, the duration, magnitude and frequency of the load current bursts are known and fixed. In these cases, the average charge current required from the LTC3108 to support the average load must be calculated, which can be easily done by the following:

$$I_{CHG} \geq I_Q + \frac{I_{BURST} \bullet t}{T}$$

Where I_Q is the sleep current on V_{OUT} required by the external circuitry in between bursts (including cap leakage), I_{BURST} is the total load current during the burst, t is

APPLICATIONS INFORMATION

the duration of the burst and T is the period of the transmit burst rate (essentially the time between bursts).

In this example, $I_Q = 5\mu A$, $I_{BURST} = 100mA$, t = 5ms and T = one hour. The average charge current required from the LTC3108 would be:

$$I_{CHG} \ge 5\mu A + \frac{100mA \bullet 0.005sec}{3600sec} = 5.14\mu A$$

Therefore, if the LTC3108 has an input voltage that allows it to supply a charge current greater than 5.14µA, the application can support 100mA bursts lasting 5ms every hour. It can be determined that the sleep current of 5µA is the dominant factor because the transmit duty cycle is so small (0.00014%). Note that for a V_{OUT} of 3.3V, the average power required by this application is only 17µW (not including converter losses).

Note that the charge current available from the LTC3108 has no effect on the sizing of the V_{OUT} capacitor (if it is assumed that the load current during a burst is much larger than the charge current), and the V_{OUT} capacitor has no effect on the maximum allowed burst rate.

TYPICAL APPLICATIONS

Peltier-Powered Energy Harvester for Remote Sensor Applications

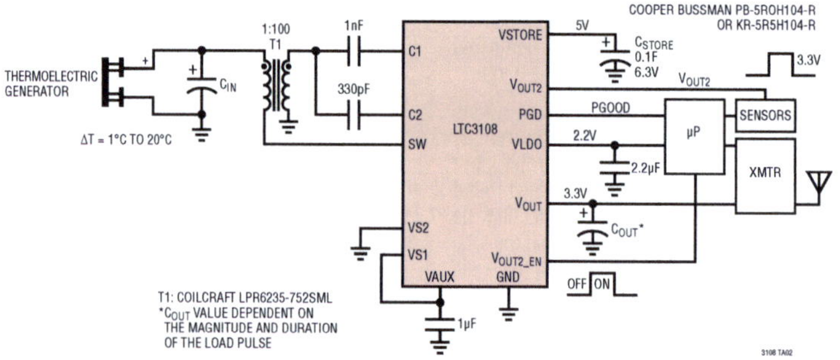

TYPICAL APPLICATIONS

Li-Ion Battery Charger and LDO Powered by a Solar Cell

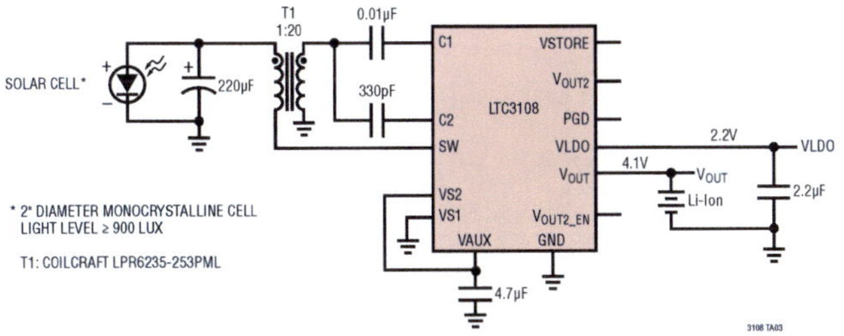

Supercapacitor Charger and LDO Powered by a Thermopile Generator

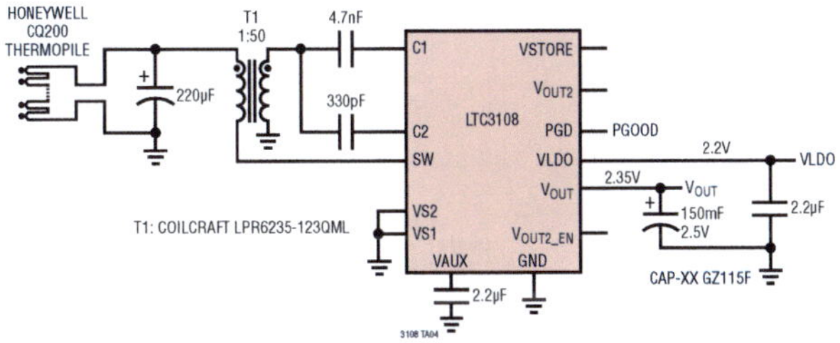

DC Input Energy Harvester and Power Manager

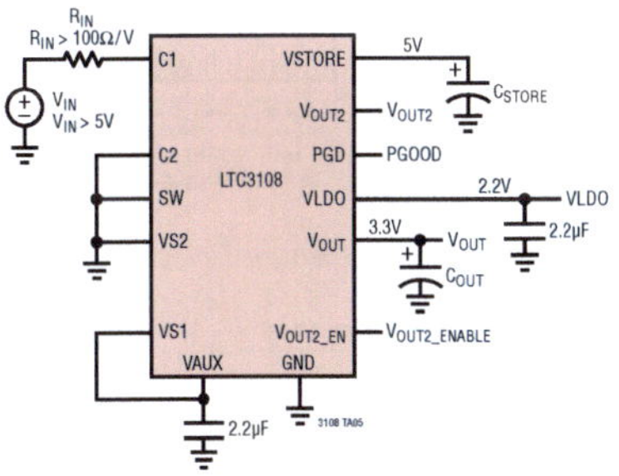

AC Input Energy Harvester and Power Manager

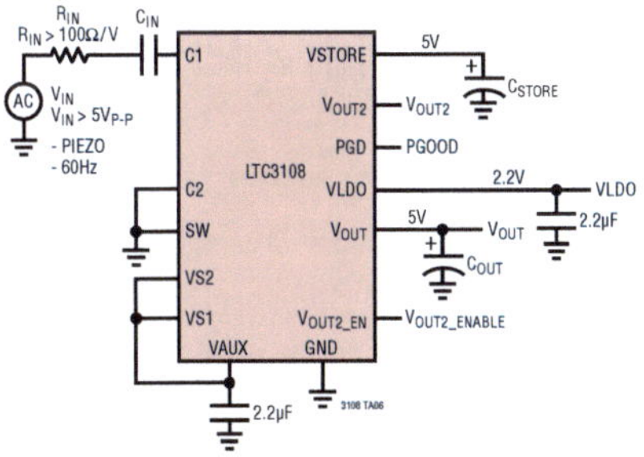

Rev. D

LTC3108

PACKAGE DESCRIPTION

GN Package
16-Lead Plastic SSOP (Narrow .150 Inch)
(Reference LTC DWG # 05-08-1641 Rev B)

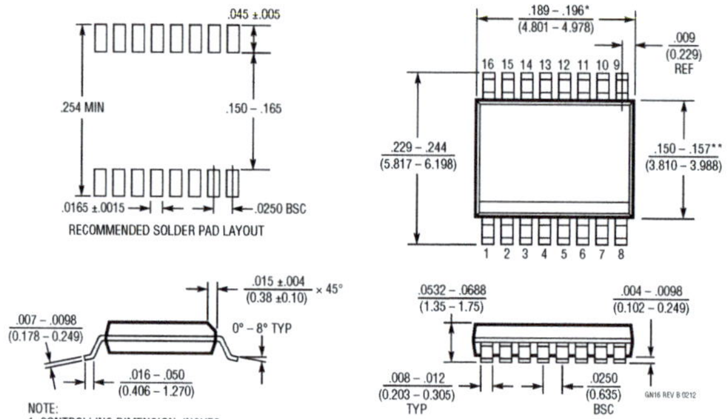

NOTE:
1. CONTROLLING DIMENSION: INCHES
2. DIMENSIONS ARE IN $\frac{\text{INCHES}}{\text{(MILLIMETERS)}}$
3. DRAWING NOT TO SCALE
4. PIN 1 CAN BE BEVEL EDGE OR A DIMPLE

*DIMENSION DOES NOT INCLUDE MOLD FLASH. MOLD FLASH SHALL NOT EXCEED 0.006" (0.152mm) PER SIDE
**DIMENSION DOES NOT INCLUDE INTERLEAD FLASH. INTERLEAD FLASH SHALL NOT EXCEED 0.010" (0.254mm) PER SIDE

DE/UE Package
12-Lead Plastic DFN (4mm × 3mm)
(Reference LTC DWG # 05-08-1695 Rev D)

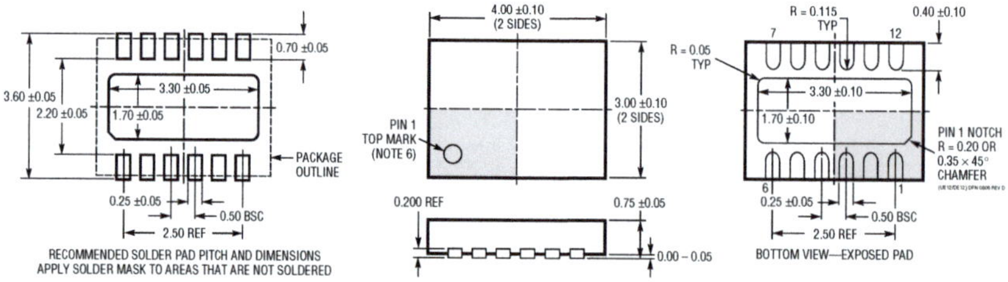

NOTE:
1. DRAWING PROPOSED TO BE A VARIATION OF VERSION (WGED) IN JEDEC PACKAGE OUTLINE M0-229
2. DRAWING NOT TO SCALE
3. ALL DIMENSIONS ARE IN MILLIMETERS
4. DIMENSIONS OF EXPOSED PAD ON BOTTOM OF PACKAGE DO NOT INCLUDE MOLD FLASH. MOLD FLASH, IF PRESENT, SHALL NOT EXCEED 0.15mm ON ANY SIDE
5. EXPOSED PAD SHALL BE SOLDER PLATED
6. SHADED AREA IS ONLY A REFERENCE FOR PIN 1 LOCATION ON THE TOP AND BOTTOM OF PACKAGE

Rev. D

20

For more information www.analog.com

REVISION HISTORY

REV	DATE	DESCRIPTION	PAGE NUMBER
A	04/10	Updated front page text and Typical Appliction	1
		Updated Absolute Maximum Ratings and Order Information sections	2
		Updated Electrical Characteristics	3
		Added graph (3108 G00) to Typical Performance Characteristics	4
		Updated Block Diagram	8
		Text added to Operation section	9
		Changes to Applications Information section	12-18
		Updated Typical Applications	18, 19, 22
		Updated Related Parts	22
B	06/13	Added vendor information to Table 5	14
C	08/13	Changed Würth transformer part numbers	14
D	03/19	Electrical Characteristics. LDO Dropout Voltage Temperature Range changed from Over Temperature to T_A = 25°C, V_{OUT}, Current Limit Max & VSTORE Current Limit Max changed from 7V to 7.5V; VAUX Clamp Voltage Max changed from 5.55V to 5.6V	3

LTC3108

TYPICAL APPLICATION

Dual TEG Energy Harvester Operates from Temperature Differentials of Either Polarity

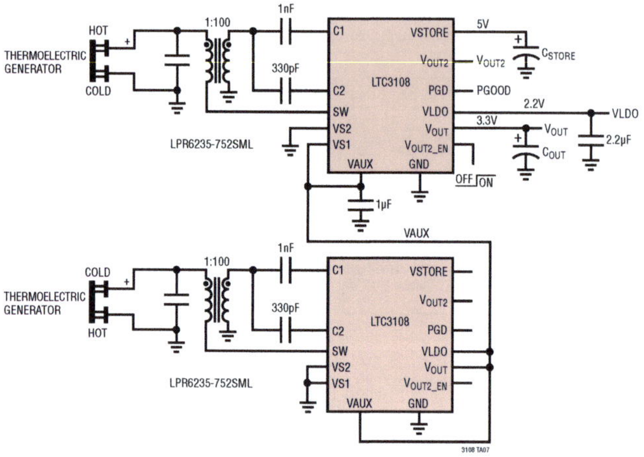

RELATED PARTS

PART NUMBER	DESCRIPTION	COMMENTS
LTC1041	Bang-Bang Controller	V_{IN}: 2.8V to 16V; I_Q = 1µA; SO-8 Package
LTC1389	Nanopower Precision Shunt Voltage Reference	$V_{OUT(MIN)}$ = 1.25V; I_Q = 0.8µA; SO-8 Package
LT1672/LT1673/ LT1674	Single-/Dual-/Quad-Precision 2µA Rail-to-Rail Op Amps	SO-8, SO-14 and MSOP-8 Packages
LT3009	3µA I_Q, 20mA Linear Regulator	V_{IN}: 1.6V to 20V; $V_{OUT(MIN)}$: 0.6V to Adj, 1.2V, 1.5V, 1.8V, 2.5V, 3.3V, 5V to Fixed; I_Q = 3µA; I_{SD} < 1µA; 2mm × 2mm DFN-8 and SC70 Packages
LTC3108-1	Ultralow Voltage Step-Up Converter and Power Manager	V_{IN}: 0.02V to 1V; V_{OUT} = 2.5V, 3V, 3.7V, 4.5V Fixed; I_Q = 6µA; 3mm × 4mm DFN-12 and SSOP-16 Packages
LTC3525L-3/ LTC3525L-3.3/ LTC3525L-5	400mA (I_{SW}), Synchronous Step-Up DC/DC Converter with Output Disconnect	V_{IN}: 0.7V to 4V; $V_{OUT(MIN)}$ = $5V_{MAX}$; I_Q = 7µA; I_{SD} < 1µA; SC70 Package
LTC3588-1	Piezoelectric Energy Generator with Integrated High Efficiency Buck Converter	V_{IN}: 2.7V to 20V; $V_{OUT(MIN)}$: Fixed to 1.8V, 2.5V, 3.3V, 3.6V; I_Q = 0.95µA; 3mm × 3mm DFN-10 and MSOP-10E Packages
LTC3642	45V, 50mA Synchronous MicroPower Buck Converter	V_{IN}: 4.5V to 45V, $60V_{MAX}$; $V_{OUT(MIN)}$: 0.8V to Adj, 3.3V Fixed, 5V Fixed; I_Q = 12µA; I_{SD} < 1µA; 3mm × 3mm DFN-8 and MSOP-8E Packages
LTC6656	850mA Precision Reference	Series Low Dropout Precision
LT8410/LT8410-1	MicroPower 25mA/8mA Low Noise Boost Converter with Integrated Schottky Diode and Output Disconnect	V_{IN}: 2.6V to 16V; $V_{OUT(MIN)}$ = $40V_{MAX}$; I_Q = 8.5µA; I_{SD} < 1µA; 2mm × 2mm DFN-8 Package
LTC4070	Micropower Shunt Li-Ion Charge	Controls Charging with µA Source

Rev. D

22

03/19

ANALOG DEVICES

www.analog.com

Index

A
ADC hat 12

B
Bandgap 25
Beer-Lambert's law 30
Betz Limit 46

C
CAD software 8
current transducer 12

D
diversion load 51
DS18B20 22

E
Electrical Metal Tubing 60
electrolyte 63
energy harvesting circuit 58

F
fall protection 34
fuel cells 55

G
Global irradiance 14

M
maximum power 27

P
photoelectric effect 25
Photovoltaic effect 25
Pi-logger 11
power coefficient 46

R
Raspberry Pi Zero 2W 8

S
Stefan-Boltzmann law 16

T
tip speed ratio 46

W
Wolfram Mathematica 18